SOLUTIONS MANUAL FOR

Environmental Chemistry

FIFTH EDITION

Colin Baird
University of Western Ontario

Michael Cann
University of Scranton

W. H. FREEMAN AND COMPANY
NEW YORK

ISBN-13: 978-1-4641-0646-0
ISBN-10: 1-4641-0646-0

Printed in the United States of America

W. H. Freeman and Company
41 Madison Avenue
New York, NY 10010
Houndmills, Basingstoke RG21 6XS England
www.whfreeman.com

Contents

Periodic Table

1	2	3	4	5	6	7	8	9	10	11	12	13	14	15	16	17	18
1 **H** 1.00797																	2 **He** 4.003
3 **Li** 6.941	4 **Be** 9.012											5 **B** 10.81	6 **C** 12.011	7 **N** 14.007	8 **O** 15.9994	9 **F** 19.00	10 **Ne** 20.179
11 **Na** 22.990	12 **Mg** 24.31											13 **Al** 26.98	14 **Si** 28.09	15 **P** 30.974	16 **S** 32.064	17 **Cl** 35.453	18 **Ar** 39.948
19 **K** 39.102	20 **Ca** 40.08	21 **Sc** 44.96	22 **Ti** 47.88	23 **V** 50.94	24 **Cr** 52.00	25 **Mn** 54.94	26 **Fe** 55.85	27 **Co** 58.93	28 **Ni** 58.69	29 **Cu** 63.55	30 **Zn** 65.38	31 **Ga** 69.72	32 **Ge** 72.63	33 **As** 74.92	34 **Se** 78.96	35 **Br** 79.90	36 **Kr** 83.80
37 **Rb** 85.47	38 **Sr** 87.62	39 **Y** 88.906	40 **Zr** 91.22	41 **Nb** 92.91	42 **Mo** 95.94	43 **Tc** (98)	44 **Ru** 101.1	45 **Rh** 102.905	46 **Pd** 106.4	47 **Ag** 107.870	48 **Cd** 112.41	49 **In** 114.82	50 **Sn** 118.69	51 **Sb** 121.75	52 **Te** 127.60	53 **I** 126.90	54 **Xe** 131.29
55 **Cs** 132.905	56 **Ba** 137.33	57–71 **Rare Earths**	72 **Hf** 178.49	73 **Ta** 180.95	74 **W** 183.85	75 **Re** 186.2	76 **Os** 190.2	77 **Ir** 192.2	78 **Pt** 195.09	79 **Au** 196.97	80 **Hg** 200.59	81 **Tl** 204.37	82 **Pb** 207.19	83 **Bi** 208.98	84 **Po** (210)	85 **At** (210)	86 **Rn** (222)
87 **Fr** (223)	88 **Ra** (226)	89–103 **Actinides**	104 **Rf** (261)	105 **Db** (260)	106 **Sg** (263)	107 **Bh** (262)	108 **Hs** (265)	109 **Mt** (266)	110 **Ds** (271)	111 **Rg** (272)	112 **Cn** (285)	113 **Uut** (286)	114 **Uuq** (289)	115 **Uup** (289)	116 **Uuh** (291)	117 **Uus** (294)	118 **Uuo** (294)

Rare Earths (Lanthanides)	57 **La** 138.91	58 **Ce** 140.12	59 **Pr** 140.91	60 **Nd** 144.24	61 **Pm** (147)	62 **Sm** 150.36	63 **Eu** 152.0	64 **Gd** 157.25	65 **Tb** 158.92	66 **Dy** 162.50	67 **Ho** 164.93	68 **Er** 167.26	69 **Tm** 168.93	70 **Yb** 173.04	71 **Lu** 174.97
Actinides	89 **Ac** 227.03	90 **Th** 232.04	91 **Pa** 231.04	92 **U** 238.03	93 **Np** 237.05	94 **Pu** (244)	95 **Am** (243)	96 **Cm** (247)	97 **Bk** (247)	98 **Cf** (251)	99 **Es** (252)	100 **Fm** (257)	101 **Md** (258)	102 **No** (259)	103 **Lr** (260)

The 1–18 group designation has been recommended by the International Union of Pure and Applied Chemistry (IUPAC).

CHAPTER 1

Stratospheric Chemistry: The Ozone Layer

Problem 1-1

From the text, $E = 119627/\lambda$ where λ is in nm and the units of the constant are nm kJ mol^{-1}

(a) $E = 119627$ nm kJ mol^{-1}/280 nm $= 427$ kJ mol^{-1}
From Figure 1-2, 280 nm lies in the UV region, at the junction between UV-B and UV-C

(b) $E = 119627$ nm kJ mol^{-1}/400 nm $= 299$ kJ mol^{-1}
Junction of UV(-A) and visible regions

(c) $E = 119627$ nm kJ mol^{-1}/750 nm $= 160$ kJ mol^{-1}
Junction of visible and infrared regions

(d) $E = 119627$ nm kJ mol^{-1}/4000 nm $= 29.9$ kJ mol^{-1}
Beginning of thermal IR region

Problem 1-2

Rearrange the formula from Problem 1-1 to obtain λ:

$\lambda = 119627/E = 119627$ nm kJ mol^{-1}/105 nm $= 1140$ nm, i.e., IR light.

Problem 1-3

Obtain E as ΔH° of the reaction

$$NO_2 \longrightarrow NO + O$$

$$\begin{aligned}\Delta H^\circ &= \Delta H_f^\circ(NO) + \Delta H_f^\circ(O) - \Delta H_f^\circ(NO_2)\\ &= 90.2 + 249.2 - 33.2\\ &= 306.2 \text{ kJ mol}^{-1}\end{aligned}$$

$\lambda = 119627/E = 119627$ nm kJ mol^{-1}/306.2 kJ mol^{-1} $= 390.7$ nm

For the complete dissociation, we need ΔH° for

$$NO_2 \longrightarrow N + 2O$$

$$\begin{aligned} \Delta H^\circ &= \Delta H_f^\circ(N) + 2\,\Delta H_f^\circ(O) - \Delta H_f^\circ(NO_2) \\ &= 472.7 + 2 \times 249.2 - 33.2 \\ &= 937.9 \end{aligned}$$

$$\lambda = 119627/E = 119627/937.9 = 127.5 \text{ nm}$$

Problem 1-4

Consider an altitude slightly lower than that at which the absolute concentration peaks. Here the number of ozone molecules is less but the number of air molecules is greater, so the relative concentration falls more than does the absolute. In contrast, just above the altitude for the peak, the ozone level has fallen but that of the air has also fallen, so the relative concentration falls less than does the absolute. From the trends, we conclude that the peak for the relative concentration will lie at a higher altitude than for the absolute.

Problem 1-5

The reaction for which we need the ΔH is

$$O_3 \longrightarrow O^* + O_2^*$$

The reactions for which we have ΔH values are

1) $O_3 \longrightarrow O + O_2$

2) $O \longrightarrow O^*$

3) $O_2 \longrightarrow O_2^*$

If we add the three reactions together, we obtain the needed reaction, so by Hess' Law,

$$\begin{aligned} \Delta H &= \Delta H_1 + \Delta H_2 + \Delta H_3 \\ &= 105 + 190 + 95 \text{ kJ mol}^{-1} \\ &= 390 \text{ kJ mol}^{-1} \end{aligned}$$

The wavelength corresponding to this energy can be obtained by rearranging the formula (see text)

$$E = 119627/\lambda$$

to solve for λ:

$$\begin{aligned} \lambda = 119627/E &= 119627 \text{ nm kJ mol}^{-1}/390 \text{ kJ mol}^{-1} \\ &= 307 \text{ nm} \end{aligned}$$

Problem 1-6

$$O_3 + NO \longrightarrow O_2 + NO_2$$
$$NO_2 \xrightarrow{\text{photon}} NO + O$$
$$O + O_2 \longrightarrow O_3$$

Net: Nil.

This sequence does *not* destroy ozone.

Problem 1-7

Since X = OH here, we obtain by substituting it for X in the two steps of Mechanism I,

$$OH + O_3 \longrightarrow HOO + O_2$$
$$HOO + O \longrightarrow OH + O_2$$

Adding these two reactions, and cancelling the common OH and HOO terms, we obtain

$$O_3 + O \longrightarrow 2\ O_2$$

Problem 1-8

$$O^* + H_2O \longrightarrow 2\ OH$$

Box 1-1, Problem 1

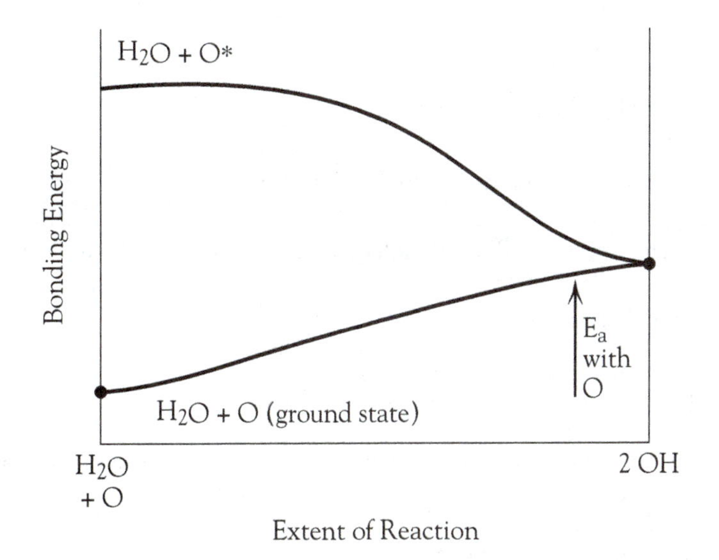

The reaction involving ground-state O is slow since $E_a > 69$ kJ mol^{-1}, but that involving O* is fast since its E_a is ~ 0.

Box 1-2, Problem 1

Given the reactants, the rate law will be

Rate = k [NO] [O_3]

Substituting the values given into the right side of the equation, then

Rate = (6.5×10^{-15} molecules^{-1} cm^3 sec^{-1}) × (5.0×10^9 molecules cm^{-3}) × (5.0×10^{12} molecules cm^{-3})

so Rate = 1.6×10^8 molecules cm^{-3} sec^{-1}

Box 1-2, Problem 2

First we need to recalculate k at the new temperature of –60°C using the Arrhenius equation:

$$k = Ae^{-E/RT}$$

From data in the Box, we have

A = 1.8×10^{-12} molecules^{-1} cm^3 sec^{-1}

E = 10,400 J mol^{-1}

T = 273 – 60 = 213 K

Thus the exponent –E/RT = –10,400 J mol^{-1} / (8.3 J K^{-1} mol^{-1} × 213 K) = –5.883

so k = 1.8×10^{-12} molecules^{-1} cm^3 sec^{-1} × $e^{-5.883}$

= 5.0×10^{-15} molecules^{-1} cm^3 sec^{-1}

Substituting this value into the same rate equation as in Problem 1, we have

Rate = 5.0×10^{-15} molecules^{-1} cm^3 sec^{-1} × (5.0×10^9 molecules cm^{-3}) × (5.0×10^{12} molecules cm^{-3})

so Rate = 1.2×10^8 molecules cm^{-3} sec^{-1}

Box 1-3, Problem 1

Labelling the reactions given in the text as 1, 2, and 3, then the expression for the rate of change of the concentration of O^* with time is

$$d[O^*]/dt = k_1[O_2] - k_2[O^*][M] - k_3[O^*][H_2O]$$

But d [O*] / dt = 0 at steady state.

Thus, setting the right-hand side of this equation to zero, and collecting terms involving O* on the left, we obtain

$$[O^*] (k_2 [M] + k_3 [H_2O]) = k_1 [O_2]$$

and thus

$$[O^*] = k_1 [O_2] / (k_2 [M] + k_3 [H_2O])$$

Box 1-3, Problem 2

Since Cl is formed in steps 1 and 3 and destroyed in step 2, we obtain

$$d [Cl] / dt = 2 k_1 [Cl_2] - k_2 [Cl] [O_3] + 2 k_3 [ClO]^2 = 0 \text{ at steady state}$$

where the coefficients of 2 are needed since two Cl's are formed, and the square of ClO is required since two of them are reactants in the step.

Similarly, $d [ClO] / dt = k_2 [Cl] [O_3] - 2 k_3 [ClO]^2 - k_4 [ClO] [NO_2] = 0$

If we add the two equations together, the middle two terms will cancel out, leaving the simple equation

$$2 k_1 [Cl_2] - k_4 [ClO] [NO_2] = 0$$

which we can use to solve for [ClO]:

$$[ClO] = 2 k_1 [Cl_2] / k_4 [NO_2]$$

Since the equation for d [Cl] / dt contains only one term for [ClO], we can substitute this expression into it to obtain an equation in [Cl]:

$$2 k_1 [Cl_2] - k_2 [Cl] [O_3] + 2 k_3 \times 4 k_1^2 [Cl_2]^2 / k_4^2 [NO_2]^2 = 0$$

The rate of destruction of O_3 is simply the middle term, k_2 [Cl] [O_3], of this equation, so by rearrangement we obtain for its net rate of change (= $-$ k_2 [Cl] [O_3]).

$$\text{Rate of change in } [O_3] = - 2 k_1 [Cl_2] - 8 k_3 k_1^2 [Cl_2]^2 / k_4^2 [NO_2]^2$$

Alternatively, the equation for [Cl] can be obtained by solving the equation for the middle term:

$$k_2 [Cl] [O_3] = 2 k_1 [Cl_2] + 8 k_3 k_1^2 [Cl_2]^2 / k_4^2 [NO_2]^2$$

$$[Cl] = 2 k_1 [Cl_2] / k_2 [O_3] + 8 k_3 k_1^2 [Cl_2]^2 / k_2 k_4^2 [O_3] [NO_2]^2$$

Box 1-3, Problem 3

Label the steps as 1, 2, and 3 in the order given in the problem.

Then, since ozone is formed in step 2 and destroyed in step 3, we have

$$d[O_3] / dt = k_2 [O] [O_2] - k_3 [NO] [O_3] = 0 \text{ at steady state.}$$

Since atomic oxygen is formed in step 1 and destroyed in step 2, we have

$$d[O] / dt = k_1 [NO_2] - k_2 [O] [O_2] = 0 \text{ at steady state.}$$

We notice that the k_2 term disappears if we add the two equations together, leaving the desired terms in NO and NO_2:

$$k_1 [NO_2] - k_3 [NO] [O_3] = 0$$

Hence, $k_1 [NO_2] = k_3 [NO] [O_3]$

And therefore the desired ratio $[NO_2] / [NO] = k_3 [O_3] / k_1$

Additional Problems

1. Let us denote the vibrational-excited oxygen molecules as O_2*. Thus the first two steps in the proposed mechanism are:

$$O_3 \xrightarrow{\lambda < 243 \text{ nm}} O_2^* + O$$

$$O_2^* + O_2 \longrightarrow O_3 + O$$

Thus the net result of these two steps is:

$$\text{net } O_2 \xrightarrow{\lambda < 243 \text{ nm}} 2\,O$$

The most likely fate of the oxygen atoms is to collide with other O_2 molecules and form ozone:

$$O + O_2 \longrightarrow O_3$$

To obtain the overall reaction, we need to include this last reaction *twice* to convert all the oxygen atoms. If we add twice this reaction to the preceding one, we obtain the overall reaction:

$$\text{net } 3\,O_2 \xrightarrow{\lambda < 243 \text{ nm}} 2\,O_3$$

(Other, but less likely fates, for the oxygen atoms are to react with each other to produce O_2, or with O_3 molecules to destroy them.)

2. Although the first step is consistent with the general one for Mechanism I, the second step is not since HOO here reacts with O_3, not with atomic oxygen.

   The sum of the two steps is

   $$OH + O_3 + HOO + O_3 \longrightarrow$$

   $$HOO + O_2 + OH + 2\,O_2$$

   After cancelling common terms from left and right sides, the overall reaction is

   $$2\,O_3 \longrightarrow 3\,O_2$$

3. The reactions as described are as follows:

   $$ClONO_2 \xrightarrow{\text{photon}} Cl + NO_3$$

   $$NO_3 \xrightarrow{\text{photon}} NO + O_2$$

   Both Cl and NO are known to destroy ozone, so presumably their reactions with O_3 follow these two:

   $$Cl + O_3 \longrightarrow ClO + O_2$$

   $$NO + O_3 \longrightarrow NO_2 + O_2$$

   To complete the cycle, we reform the $ClONO_2$ by the known combination reaction of ClO with NO_2:

   $$ClO + NO_2 \longrightarrow ClONO_2$$

   Summing all five reactions, we obtain:

   $$2\,O_3 \xrightarrow{\text{photons}} 3\,O_2$$

4. Since the first step destroys ozone, subsequent steps must reform it because the overall reaction is a null cycle. The only ozone creation reaction we know is the combination of atomic oxygen with O_2:

   $$O + O_2 \longrightarrow O_3$$

   Thus the mechanism must involve other steps that *create* atomic oxygen from XO. Since mechanism II is involved, two XO units must collide; one such reaction that produces atomic oxygen will be:

   $$XOOX \longrightarrow 2\,X + 2\,O$$

Thus the overall sequence would be

$$2\,[X + O_3 \longrightarrow XO + O_2]$$
$$2\,XO \longrightarrow XOOX$$
$$XOOX \longrightarrow 2\,X + 2\,O$$
$$2\,[O + O_2 \longrightarrow O_3]$$

net null

5. The process of concern transforms diatomic chlorine gas, Cl_2, into atomic chlorine and can be written as:

$$Cl_2\,(g) \longrightarrow 2\,Cl\,(g)$$

Thus, $\Delta H = 2\Delta H_f\,(Cl, g) - \Delta H_f\,(Cl_2, g)$

We are given $\Delta H_f\,(Cl, g) = +121.7$ kJ mole^{-1}, and since Cl_2 (g) is stated to be the stablest form of the element, by convention in thermochemistry, $\Delta H_f\,(Cl_2, g) = 0$.

$$\text{Thus, } \Delta H = 2 \times 121.7 - 0$$
$$= 243.4 \text{ kJ mol}^{-1}$$

To convert to wavelength, we use the formula:

$$E = 119627/\lambda$$

$$\text{so } \lambda = 119627/E = 119627 \text{ nm kJ mol}^{-1}/243.4 \text{ kJ mol}^{-1}$$
$$= 491 \text{ nm}$$

This wavelength lies in the visible region.

6. (a) If the NO_2 concentration is increased, the rate of its reaction with HOO will increase, thereby decreasing the HOO concentration in the lower stratosphere. Since HOO is required to complete the catalyst ozone destruction by the OH/HOO cycle, it will be slowed and the rate of ozone destruction by them would decrease.

(b) In the mid-stratosphere and high-stratosphere, NO_2 would return to NO by reaction with atomic oxygen, and the NO returns to NO_2 by destruction of ozone:

$$NO_2 + O \longrightarrow NO + O_2$$
$$NO + O_3 \longrightarrow NO_2 + O_2$$

$$\text{net } O_3 + O \longrightarrow 2\,O_2$$

Thus any increase in the NO_2 concentration would increase the rate of ozone destruction at these levels.

(c) Supersonic airplanes that emit substantial amounts of nitrogen oxides should fly in the lower atmosphere only, because they would destroy ozone at higher levels in the stratosphere.

7. Recall that the rate for a reaction step equals the rate constant k times the concentration of each reactant molecule. Thus for the reaction between O^* and CH_4, the rate is

$$\begin{aligned} \text{Rate} &= k\,[O^*]\,[CH_4] \\ &= 3 \times 10^{-10}\ \text{cm}^3\ \text{molecules}^{-1}\ \text{s}^{-1} \times 100\ \text{molecules cm}^{-3} \times 10^{11}\ \text{molecules cm}^{-3} \\ &= 3 \times 10^{3}\ \text{molecules cm}^{-3}\ \text{s}^{-1} \end{aligned}$$

To deduce the destruction over one year, we deduce the number of seconds in 1 year:

$$\frac{60\ \text{seconds}}{1\ \text{minute}} \times \frac{60\ \text{minutes}}{1\ \text{hour}} \times \frac{24\ \text{hours}}{1\ \text{day}} \times \frac{365\ \text{days}}{1\ \text{year}} = 3.15 \times 10^{7}\ \text{seconds/year}$$

Thus the rate of methane destruction per year is:

$$\frac{3 \times 10^{3}\,\text{molecules cm}^{-3}}{\text{second}} \times \frac{3.15 \times 10^{7}\,\text{seconds}}{\text{year}}$$

$$= 9.46 \times 10^{10}\ \text{molecules cm}^{-3}\ \text{year}^{-1}$$

Finally, to convert to grams of methane, we convert molecules to grams:

$$1\ \text{mole}\ CH_4 = 6.02 \times 10^{23}\ \text{molecules}\ CH_4 = 16.0\ \text{grams}\ CH_4$$

Thus

$$9.46 \times 10^{10}\,\text{molecules} \times \frac{16.0\ \text{g}\ CH_4}{6.02 \times 10^{23}\,\text{molecules}\ CH_4}$$

$$= 2.52 \times 10^{-12}\ \text{g}\ CH_4$$

Thus, in every cubic centimeter, a total of 2.52×10^{-12} grams of methane are destroyed per year.

8. The reactions of the two catalysts with atomic oxygen are:

$$Cl + O_3 \longrightarrow ClO + O_2 \qquad \ldots(1)$$

and $$OH + O_3 \longrightarrow HOO + O_2 \qquad \ldots(2)$$

The rate laws for these reactions will each be first-order in ozone and in catalyst, because 1 molecule of each is involved.

$$\text{Rate}\ (1) = k_1\,[Cl]\,[O_3]$$

$$\text{Rate}\ (2) = k_2\,[OH]\,[O_3]$$

The ratio of rates is then given by:

$$\frac{\text{Rate (1)}}{\text{Rate (2)}} = \frac{k_1}{k_2}\frac{[\text{Cl}]}{[\text{OH}]}$$

Given the information that [OH] = 100 [Cl], thus

$$\frac{\text{Rate (1)}}{\text{Rate (2)}} = \frac{k_1}{k_2}\frac{[\text{Cl}]}{100\,[\text{Cl}]} = \frac{k_1}{100\,k_2}$$

$$\text{Now } \frac{k_1}{k_2} = \frac{3 \times 10^{-11}\, e^{-250/T}}{2 \times 10^{-12}\, e^{-940/T}} = 15\, e^{+690/T}$$

Substituting T = 273 – 50 = 223K, thus

$$\frac{k_1}{k_2} = 15\, e^{690/223} = 330$$

$$\text{Thus } \frac{\text{Rate (1)}}{\text{Rate (2)}} = \frac{1}{100} \times 330 = 3.3$$

The rate of ozone destruction in the ozone hole conditions described will be given by:

$$\begin{aligned}\text{Rate (1)} &= k_1\, [\text{Cl}]\, [\text{O}_3] \\ &= 3 \times 10^{-11}\, e^{-250/T} \times 4 \times 10^5 \times 2 \times 10^{12} \\ &= 3 \times 10^{-11}\, e^{-250/193} \times 4 \times 10^5 \times 2 \times 10^{12} \\ &= 7 \times 10^6 \text{ molecules cm}^{-3}\text{ sec}^{-1}\end{aligned}$$

9. The Arrhenius equation is

$$k = Ae^{-E/RT}$$

The ratio of rate constants for the two reactions is therefore:

$$\frac{k_1}{k_2} = \frac{A_1}{A_2}\frac{e^{-E_1/RT}}{e^{-E_2/RT}}$$

but, because we are given that $A_1 = A_2$, therefore

$$\frac{k_1}{k_2} = e^{-(E_1-E_2)/RT}$$

$$\begin{aligned}\text{Now}\quad E_1 &= 30\text{ kJ mol}^{-1} = 30{,}000\text{ J mol}^{-1} \\ \text{and}\quad E_2 &= 3\text{ kJ mol}^{-1} = 3{,}000\text{ J mol}^{-1} \\ \text{Since}\quad R &= 8.314\text{ J K}^{-1}\text{ mol}^{-1}\text{, and} \\ T &= -30 + 273 = 243\text{ K,} \\ \text{therefore}\quad \frac{k_1}{k_2} &= e^{-(30{,}000\,-\,3{,}000)/8.314\,\times\,243} \\ &= 1.5 \times 10^{-6}\end{aligned}$$

Since we are told that the reactant concentrations are equal, their values cancel from the ratio of the rate equations, and the ratio of rates is equal to that of the rate constants, namely 1.5×10^{-6}.

CHAPTER 2

The Ozone Holes

Problem 2-1

$$O_3 + Cl \longrightarrow ClO + O_2$$
$$O_3 + Br \longrightarrow BrO + O_2$$
$$ClO + BrO \longrightarrow Cl + Br + O_2$$

$$\text{overall } 2\,O_3 \longrightarrow 3\,O_2$$

Problem 2-2

The Cl_2O_2 mechanism will become more dominant.

Since $\text{rate}_{Cl_2O_2} = k\,[ClO]^2$
whereas $\text{rate}_{ClOBrO} = k'\,[ClO]\,[BrO]$

Thus, $\text{rate}_{Cl_2O_2}/\text{rate}_{ClOBrO} = \dfrac{k\,[ClO]}{k'\,[BrO]}$

In other words, because the rate ratio is proportional to the chlorine/bromine ratio in the stratosphere, as this ratio increases, the Cl_2O_2 rate relative to the ClO/BrO rate will increase, and become even more important.

Problem 2-3

In the upper stratosphere, atomic O is relatively plentiful and ClO is normally rare, so the rate for the ClO + O process that depends upon the first power of the ClO concentration will be much greater than one that depends on the square of the ClO concentration.

Problem 2-4

If C—F bonds are stronger than O—F, then the fluorine abstraction reaction

$$OH + CF_2Cl_2 \longrightarrow HOF + CFCl_2$$

will be *endothermic* and therefore have a high activation energy and thus be *slow*. Thus, the reaction will not be a realistic sink for CFCs.

Problem 2-5

(a) Mechanism I

$$F + O_3 \longrightarrow OF + O_2$$
$$OF + O \longrightarrow F + O_2$$

Mechanism II

$$F + O_3 \longrightarrow OF + O_2$$
$$2\,OF \longrightarrow [FOOF] \longrightarrow 2\,F + O_2$$

(b) Mechanism I here is changed to

$$F + O_3 \longrightarrow FO + O_2$$
$$FO + O_3 \longrightarrow F + 2\,O_2$$

$$\text{net } 2\,O_3 \longrightarrow 3\,O_2$$

Problem 2-6

For OH, Mechanism I of catalytic ozone destruction is:

$$OH + O_3 \longrightarrow HOO + O_2$$
$$HOO + O \longrightarrow OH + O_2$$

If $CF_3{-}O$ replaces $H{-}O$, this mechanism would be:

$$CF_3O + O_3 \longrightarrow CF_3OO + O_2$$
$$CF_3OO + O \longrightarrow CF_3O + O_2$$

Problem 2-7

If the C—H bond energy in propane and butane is less than in methane, then the inactivation of atomic chlorine by the reactions

$$Cl + C_3H_8 \longrightarrow HCl + C_3H_7$$

and $Cl + C_4H_{10} \longrightarrow HCl + C_4H_9$

will have lower activation energies and thus faster rates than for

$$Cl + CH_4 \longrightarrow HCl + CH_3.$$

Thus because of the faster propane and butane reactions, the steady-state concentration of atomic Cl will decrease more than with methane and less ozone will be destroyed due to the lower concentration of active Cl.

Problem 2-8

Presumably CH_3Cl, CH_2Cl, and $CHCl_3$ are destroyed to a large extent (by the H abstraction according to our principles) in the troposphere since they are not candidates for regulation. Thus most of the chlorine in these molecules will find a sink in the troposphere and not have time to rise to the stratosphere to destroy ozone.

Green Chemistry Problems

1. (a) Principle 2. Alternative reaction conditions for green chemistry.

   (b) Principle 1. Prevention of waste.

   Principle 4. Preserving efficacy of function while reducing environmental damage.

2. The use of carbon dioxide as a blowing agent uses waste products from other processes. Unlike CFCs, carbon dioxide does not deplete the ozone layer. Unlike hydrocarbons, carbon dioxide does not contribute to the pollution of the troposphere.

3. Although carbon dioxide is a greenhouse gas and thus contributes to global warming, the carbon dioxide that is used as a blowing agent is captured as a waste byproduct from processes such as natural gas production and ammonia production. This waste carbon dioxide would have normally been vented to the atmosphere.

4. (a) Principle 3. The design chemicals that are less toxic than current alternatives or inherently safer with regard to accident potential.

   (b) Principle 1. Prevention of waste. Much less harpin is used compared to conventional pesticides.

   Principle 4. Preserving efficacy of function while reducing toxicity. Lower toxicity (lowest EPA toxicity category).

   Principle 6. Energy reduction. Harpin is produced by natural fermentation, not from petrochemical feedstocks.

   Principle 10. Degradation in the environment.

5. Harpin does not effect the pest directly. It is applied to the plant, which elicits the plant's own natural defenses.

6. Harpin is readily degraded in the environment by microorganisms and UV light.

Additional Problems

1. (a) For a gas, the number of moles in any sample is proportional to the sample's volume, because $PV = nRT$. The volume here equals the height of the layer, when converted to pure ozone at 1 atmosphere and 0°C, times the surface area. Thus moles $\propto$ height.

   As defined in the text, 1 Dobson Unit equals 0.01 mm or 0.001 cm of ozone at 1 atmosphere pressure. Since PV is a constant for a gas sample, the product 1 atm $\times$ 0.001 cm ($\times$ surface area) can be re-expressed as 0.001 atm $\times$ 1 cm ($\times$ surface area); in other words, as a milli-atmosphere (0.001 atm) times a centimeter.

   Thus, 1 DU = 0.001 cm atm = 1 matm cm.

   (b) We are told that the volume of a sphere is proportional to the cube of its radius. The two spheres that are involved here are the Earth, radius r, and the Earth plus the 3.5 mm of air-space above its surface that corresponds to pure ozone when the column ozone is collapsed to standard temperature and pressure. The volume of ozone is the difference in volume between these spheres:

   Volume of ozone = Volume of sphere of Earth + ozone
   − Volume of sphere of Earth alone

   $$= \frac{4}{3}\pi(r + \Delta)^3 - \frac{4}{3}\pi r^3 = \frac{4}{3}\pi[(r + \Delta)^3 - r^3]$$

   where r is the Earth's radius and Δ is 3.5 mm.

   Now expanding

   $$(r + \Delta)^3 = r^3 + 3r^2\Delta + 3r\Delta^2 + \Delta^3$$

   so the term in square brackets reduces to

   $$3r^2\Delta + 3r\Delta^2 + \Delta^3$$

   Since $r >> \Delta$, the lead term, $3r^2\Delta$, will completely dominate.

   $$\text{Thus, Volume of ozone} = \frac{4}{3}\pi(3r^2\Delta) = 4\pi r^2\Delta$$

   Now $r = 6400 \text{ km} = 6400 \times 10^3 \text{ m}$
   and $\Delta = 3.5 \text{ mm} = 3.5 \times 10^{-3} \text{ m}$

   $$\begin{aligned}\text{Thus, Volume} &= 4 \times 3.14 \times (6400 \times 10^3 \text{ m})^2 \times (3.5 \times 10^{-3} \text{ m})\\ &= 1.8 \times 10^{12} \text{ m}^3\\ &= 1.8 \times 10^{15} \text{ L}\end{aligned}$$

   since $1 \text{ m}^3 = 1000 \text{ L}$

The mass of ozone can be deduced from the ideal gas law by first deducing its number of moles:

$$n = PV/RT = 1.0 \text{ atm} \times 1.8 \times 10^{15} \text{ L}/0.082 \text{ L atm mol}^{-1} \text{ K}^{-1} \times 273 \text{ K}$$
$$= 8.0 \times 10^{13} \text{ moles}$$

Since each mole of O_3 has a mass of 48 grams, the total mass of ozone is $48 \times 8.0 \times 10^{13} = 4 \times 10^{15}$ grams

2. To obtain the formula, add 90 to the code number to give the three-digit integer corresponding to the number of C, H, and F atoms, respectively. Then deduce the number of chlorine atoms present.

   (a) $12 + 90 = 102$, so this corresponds to $C_1H_0F_2$; because the carbon atom forms a total of four bonds, we deduce that two chlorines must be present. Therefore, the formula is CF_2Cl_2.

   (b) $113 + 90 = 203$, which gives $C_2H_0F_3$. Total hydrogens + substituents in alkane derivatives is $2n + 2$ if there are n carbons, so there must be $2 \times 2 + 2 = 6$ atoms total bonded to the carbons here. Thus, there are 3 Cl atoms because there are only three (H + F) atoms. The formula must be $C_2F_3Cl_3$.

   (c) $123 + 90 = 213$, which gives $C_2H_1F_3$. Since, as in (b), six atoms are bonded to the carbons, there must be $6 - 1 - 3 = 2$ chlorine atoms; thus the formula is $C_2HF_3Cl_2$.

   (d) $134 + 90 = 224$, which gives $C_2H_2F_4$. As in (b), 6 atoms must be attached to two carbons, but $2 + 4 = 6$, so no chlorine atoms are present. The formula is $C_2H_2F_4$.

3. Write the formulas as $C_nH_mF_p$, ignoring the chlorines, so the code number is (nmp − 90).

   (a) Formula rewritten is $C_2H_3F_0$, so code is $230 - 90 = 140$.

   (b) Formula rewritten is $C_1H_0F_0$, so code is $100 - 90 = 10$.

   (c) Formula rewritten is $C_2H_3F_1$, so code is $231 - 90 = 141$.

4. To the code of 134, we add 90, giving 224. Thus the numbers of C, H, and F atoms should be 2, 2, and 4 respectively, yielding the formula $C_2H_2F_4$, which indeed is consistent with the formula CH_2FCF_3. Since this compound has an isomer, namely CHF_2CHF_2, there must be an additional designation—a or b—used to distinguish them.

   For CH_2F_2, as explained in the solution to Problem 3, we have n = 1, m = 2, and p = 2, so its code number is 122 – 90 = 32. Similarly, for CHF_2CF_3, we have n = 2, m = 1, and p = 4, so its code number is 214 – 90 = 124.

   No, 410 is not the code number for either of the components of R-410a.

5. The conditions under which the chlorine dimer mechanism is important are:

   (a) chlorine is activated on the surface of particles
   (b) the Cl_2O_2 dimer is stable to thermal decomposition
   (c) most NO_2 is tied up as HNO_3

   At mid-latitudes, the temperature in the lower stratosphere would not be very low, so the Cl_2O_2 dimer would probably ultimately thermally dissociate back to ClO before photolysis could occur:

   $$Cl_2O_2 \longrightarrow 2\ ClO$$

   Thus the cycle of Cl $\longrightarrow$ ClO $\longrightarrow$ Cl would not occur, and ozone would not be catalytically destroyed.

   In addition, nitrogen dioxide would not be inactivated but available to convert ClO immediately back to the inactive $ClONO_2$ form:

   $$ClO + NO_2 \longrightarrow ClONO_2$$

6. The reaction step of interest is:

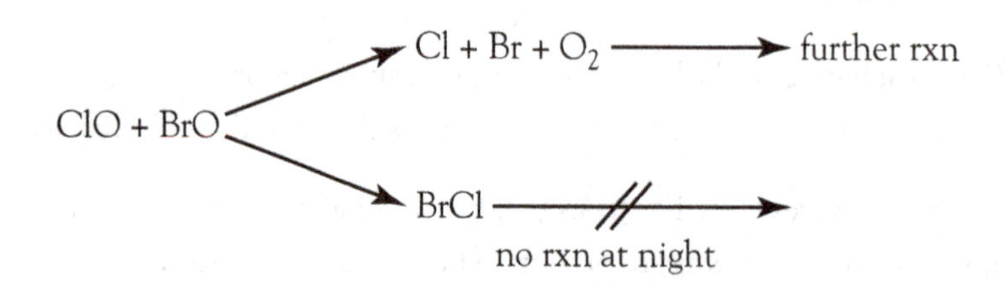

   At any given time, most reactions of ClO and BrO will produce Cl and Br, which will continue the cycle and soon again appear as ClO and BrO. However, in each cycle, a fraction of the halogen will enter the BrCl sink and remain as such until dawn. Thus the fraction of halogen existing as BrCl will increase as time goes on and the cycle is repeated over and over, each time with less and less halogen involved.

7. See text for this information.

8. The advantage of using hydrocarbons rather than HFCs or HCFCs is their short atmospheric lifetime and their lack of chlorine content (in contrast to HCFCs). Their main disadvantage is their flammability. A flame retardant should be added to hydrocarbons to make them safer.

9. CHF_3 contains no chlorine, only H and F, so it has zero ozone-depleting potential.

   $CHFCl_2$ contains chlorine, but also contains H, making it an HCFC. H atom abstraction via hydroxyl radical attack thus provides a significant sink for this compound in the troposphere, so only a fraction of this molecule released at ground level will make it to the stratosphere, where its chlorine will eventually be released and it will destroy ozone.

   CF_3Cl and $CFCl_3$ are both CFCs with no tropospheric sinks, and thus most of the mass of each of them will rise to the stratosphere and catalytically destroy ozone there. Of the two, $CFCl_3$ has 3 Cl per molecule compared to 1 in the case of CF_3Cl, and so $CFCl_3$ would destroy more ozone.

   Thus, the order in terms of increasing ODP is: $CHF_3 < CHFCl_2 < CF_3Cl < CFCl_3$.

CHAPTER 3

The Chemistry of Ground-Level Air Pollution

Problem 3-1

Original concentration is:

$$\frac{8.7 \times 10^{6}\text{ molecules OH}}{1\text{ cm}^{3}\text{ of air}}$$

For a molar concentration, i.e., moles of OH per liter of air, we must convert molecules of OH to moles, and 1 cm^3 air to liters.

$$8.7 \times 10^{6}\text{ molecules OH} \times \frac{1\text{ mole OH}}{6.02 \times 10^{23}\text{ molecules OH}} = 1.4 \times 10^{-17}\text{ moles OH}$$

Thus, the molar concentration is:

$$1.4 \times 10^{-17}\text{ moles OH} / 10^{-3}\text{ L air}$$

$$= 1.4 \times 10^{-14}\text{ moles / L}$$

To obtain the concentration in ppt, we must deduce the number of trillions of molecules in 1 cm^3 of air, because the numerator of the original concentrations is already in numbers of molecules.

For the 1 cm^3 of air sample,

$$n = PV/RT = 1.0\text{ atm} \times 10^{-3}\text{ L}/0.082\text{ L atm mol}^{-1}\text{ K}^{-1} \times 288\text{ K}$$
$$= 4.2 \times 10^{-5}\text{ moles air}$$

$$4.2 \times 10^{-5}\text{ moles air} \times \frac{6.02 \times 10^{23}\text{ molecules air}}{1\text{ mole air}}$$

$$= 2.5 \times 10^{19}\text{ molecules of air}$$
$$= 2.5 \times 10^{7} \times (10^{12}\text{ molecules of air}) = 2.5 \times 10^{7}\text{ trillion molecules}$$

Concentration is 8.7×10^{6} molecules OH / 2.5×10^{7} trillion molecules of air, or 0.35 ppt.

Problem 3-2

The three reactions are:

$$NO_2 + \text{UV-A} \rightarrow NO + O$$

$$O + O_2 \rightarrow O_3$$

$$O_3 + NO \rightarrow O_2 + NO_2$$

Adding the left sides together, and the right sides, gives

$$NO_2 + \text{UV-A} + O + O_2 + O_3 + NO \rightarrow NO + O + O_3 + O_2 + NO_2$$

There remain no chemicals after formulas common to both sides are cancelled; that is, it is a null reaction.

Problem 3-3

Since the standards are listed in terms of mass per volume, the simplest conversion would be to interpret the ppb units as volume / volume. Thus we convert the masses listed in the denominators of the standards into volumes (at a standard pressure of 1 atm) using the ideal gas equation. First convert the mass of ozone into moles of it:

$$120 \times 10^{-6} \text{ g } O_3 \times (1 \text{ mol } O_3 / 48.0 \text{ g } O_3) = 2.5 \times 10^{-6} \text{ mol } O_3$$

Now PV = nRT, so

$$V = nRT / P = 2.5 \times 10^{-6} \text{ mol} \times 0.0821 \text{ L atm mol}^{-1}\text{K}^{-1} \times (273 + 27) / 1 \text{ atm}$$

$$\text{So } V = 6.16 \times 10^{-5} \text{ L}$$

Using the fact that 1 L = 0.001 m^3, we find the volume of ozone is 6.16×10^{-8} m^3. Thus the concentration of ozone is 6.16×10^{-8} m^3 O_3 / 1 m^3 air, or 61.6 m^3 O_3 / 10^9 m^3 air, so its concentration in the new units is 61.6 ppb.

Similarly, the 100 μg m^{-3} standard is equal to 51.3 ppb.

Problem 3-4

From Problem 3-3, 120 μg of ozone is 2.5×10^{-6} moles of it.

The U.S. standard is equivalent to 75 moles ozone in10^9 moles of air.

Thus in their regulations, the ratio of ozone quoted is

U.S. moles / EU moles = $75 / 2.5 \times 10^{-6} = 3.0 \times 10^{7}$

To be equivalent concentrations, the ratio of air volumes in the two standards must equal this mole ratio. Thus

$V_{US} / V_{EU} = 3.0 \times 10^{7}$

But $V_{EU} = 1\ m^3 = 1000$ L, so $V_{US} = 1000 \times 3.0 \times 10^{7} = 3.0 \times 10^{10}$ L

The temperature at which 10^9 moles of air occupy 3.0×10^{10} L can be deduced from the Ideal Gas Law:

PV = nRT,

so T = PV / nR = 1.0 atm $\times$ 2.0×10^{10} L / 10^9 moles $\times$ 0.082 L atm mole^{-1} K^{-1}

= 366 K

The two standards become equivalent at 366 K, that is, at 93°C.

Problem 3-5

In pencil, draw a horizontal line on Figure 3-8 starting at the 0.20 point on the vertical axis. Then draw vertical straight lines starting at both 0.40 and 0.50 on the horizontal axis. Notice that the 0.5 line for VOC and the 0.2 line for NOx meet at the 160 ppm ozone curve, and that the 0.4 VOC line meets the 0.2 NOx line at the 80 ppm ozone curve. Therefore, reducing [VOC] from 0.5 to 0.4 (while maintaining [NOx] at 0.20) will cut ozone in half, from 160 to 80 ppm. Because VOC level changes have such an effect in this region, we can conclude it is VOC-limited.

Problem 3-6

With NO_x = 0.20, the ozone concentration will be slightly less than 160 ppb since the 0.2 and 0.5 ppm point lies slightly to the left of the 160 ppb contour.

With NO_x = 0.08, the ozone concentration will be greater 160 ppb since the 0.08 and 0.5 ppm point lies to the right of the 160 ppb contour.

Thus, in this region of the graph, lowering the NO_x concentration actually increases the concentration of ozone. As explained in the text, this occurs because lowering the concentration of NO_2 leads to less OH being consumed in their reaction together producing nitric acid, and hence more OH is available to initiate reaction with the VOCs present.

Problem 3-7

(a) The unbalanced reaction is

$$NO + CO \longrightarrow N_2 + CO_2$$

The O from each NO molecule will be combined with a CO_2 molecule, so equal numbers of NO and CO are required. After balancing by inspection, we obtain

$$2\ NO + 2\ CO \longrightarrow N_2 + 2\ CO_2$$

(b) The unbalanced equation is

$$NO + C_6H_{14} \longrightarrow N_2 + CO_2 + H_2O$$

Every C_6H_{14} molecule will oxidize to give 6 CO_2 and 7 H_2O molecules, therefore requiring 19 O atoms, one from each of 19 NO molecules. The balanced equation becomes

$$38\ NO + 2\ C_6H_{14} \longrightarrow 19\ N_2 + 12\ CO_2 + 14\ H_2O$$

Problem 3-8

The unbalanced equation is

$$NO + CO(NH_2)_2 \longrightarrow N_2 + CO_2 + H_2O$$

Assuming the carbon in urea has a +4 oxidation number, that of each nitrogen atom in the compound is –3, for a total of –6. Since the N atom in NO is +2, three times as much NO as urea will be required to react together to produce N_2, in which the N oxidation number is zero. The balanced equation is

$$6\ NO + 2\ CO(NH_2)_2 \longrightarrow 5\ N_2 + 2\ CO_2 + 4\ H_2O$$

Problem 3-9

The equation can be balanced most readily by realizing that in NH_3 the oxidation number of nitrogen is –3, in NO_2 it is +4, and in N_2 it is zero; thus to conserve charge, the number of NH_3 molecules must be 4/3 times the number of NO_2 molecules. Thus the initially unbalanced equation:

$$NH_3 + NO_2 \longrightarrow N_2 + H_2O$$

becomes $4/3\ NH_3 + NO_2 \longrightarrow 7/6\ N_2 + 2\ H_2O$

Multiplying by 6 to obtain all integers, we obtain:

$$8\,NH_3 + 6\,NO_2 \longrightarrow 7\,N_2 + 12\,H_2O$$

Now 10 ppm NO_2 means 10 L of NO_2 in 10^6 L of air, so the volume of NO_2 is:

$$1000 \text{ L air} \times \frac{10 \text{ L } NO_2}{10^6 \text{ L air}} = 0.010 \text{ L } NO_2$$

Since coefficients in reaction equations can be interpreted as volumes that react with each other (Avogadro's Law of Combining Volumes), we can convert this volume of NO_2 to that of NH_3:

$$0.010 \text{ L } NO_2 \times \frac{8 \text{ L } NH_3}{6 \text{ L } NO_2} = 0.0133 \text{ L } NH_3$$

From the ideal gas equation, we can solve for the moles of NH_3:

$$\begin{aligned} n &= PV/RT = 1.0 \text{ atm} \times 0.0133 \text{ L}/0.082 \text{ L atm mol}^{-1} \text{ K}^{-1} \times 300 \text{ K} \\ &= 0.00054 \text{ moles } NH_3 \end{aligned}$$

Since 1 mole NH_3 has a mass of 17.0 g, the mass of NH_3 here is $0.00054 \times 17.0 = 0.0092$ grams.

Problem 3-10

First deduce the mass of SO_2 produced by burning the coal; the reaction is:

$$S + O_2 \longrightarrow SO_2$$

$$1000 \text{ kg coal} \times \frac{0.05 \text{ kg S}}{1 \text{ kg coal}} \times \frac{64.1 \text{ kg } SO_2}{32.1 \text{ kg S}} = 99.8 \text{ kg } SO_2$$

where we have used the fact that the molar mass of SO_2 is 64.1 and that of sulfur is 32.1.

The reaction with $CaCO_3$ of the SO_2 is (see text):

$$CaCO_3 + SO_2 \longrightarrow CaSO_3 + CO_2$$

Since the molar mass of $CaCO_3$ is 100.1, we can convert the mass of SO_2, to that of $CaCO_3$:

$$99.8 \text{ kg } SO_2 \times \frac{100.1 \text{ kg } CaCO_3}{64.1 \text{ kg } SO_2} = 156 \text{ kg } CaCO_3$$

Problem 3-11

From the description, the unbalanced equation is:

$$NaOH + SO_2 \longrightarrow H_2O + Na_2SO_3$$

This is readily balanced by balancing the Na atoms to obtain:

$$2\,NaOH + SO_2 \longrightarrow H_2O + Na_2SO_3$$

Calcium sulfite contains Ca^{2+} and SO_3^{2-} and so its formula must be $CaSO_3$. The skeleton equation for the process described in the problem is:

$$Na_2SO_3 + \underline{\ ?\ } \longrightarrow CaSO_3 + NaOH$$

The elements missing from the reactant side are clearly Ca and H and perhaps oxygen; these could be supplied conveniently by $Ca(OH)_2$. Thus we obtain the unbalanced equation:

$$Na_2SO_3 + Ca(OH)_2 \longrightarrow CaSO_3 + NaOH$$

Balancing the Na and the H gives the final balanced equation:

$$Na_2SO_3 + Ca(OH)_2 \longrightarrow CaSO_3 + 2\ NaOH$$

Adding the equation to the previous balanced one (that produced Na_2SO_3) gives the net reaction once common terms are canceled:

$$\begin{array}{lll} & 2\ NaOH + SO_2 \longrightarrow & H_2O + Na_2SO_3 \\ & Na_2SO_3 + Ca(OH)_2 \longrightarrow & CaSO_3 + 2\ NaOH \\ \hline \text{net} & SO_2 + Ca(OH)_2 \longrightarrow & CaSO_3 + H_2O \end{array}$$

Problem 3-12

For the process $HSO_3^- \rightleftharpoons H^+ + SO_3^{2-}$

we have $K_a = 1.2 \times 10^{-7} = [H^+]\,[SO_3^{2-}]\,/\,[HSO_3^-]$

But due to the previous equilibrium, we know $[HSO_3^-] = [H^+]$

Thus these terms cancel from the K_a equation, and we have as our result

$$[SO_3^{2-}] = 1.2 \times 10^{-7}$$

Problem 3-13

Since $K_H = [H_2SO_3]\,/\,P$

and $K_H = 1.0\ M\ atm^{-1}$

$P = 1.0 \times 10^{-6}\ atm$

Thus $[H_2SO_3] = K_H P = 1.0\ M\ atm^{-1} \times 10^{-6}\ atm = 1.0 \times 10^{-6}\ M$

Now $H_2SO_3 \rightleftharpoons H^+ + HSO_3^-$

so $K_a = [H^+]\,[HSO_3^-]\,/\,[H_2SO_3]$

or $[H^+]\,[HSO_3^-] = K_a\,[H_2SO_3]$

$= 1.7 \times 10^{-2} \times 10^{-6}$

$= 1.7 \times 10^{-8}$

Now $[HSO_3^-] = [H^+]$,
so $[H^+]^2 = 1.7 \times 10^{-8}$
or $[H^+] = 1.3 \times 10^{-4}$ M
i.e. pH = 3.88

Problem 3-14

As in Problem 3-13, let us assume

$[HSO_3^-] = [H^+] = 1.0 \times 10^{-4}$
since pH = 4.0.

Since $K_a = [H^+]\,[HSO_3^-] / [H_2SO_3]$
then $[H_2SO_3] = [H^+]\,[HSO_3^-] / K_a$
$= (1.0 \times 10^{-4})^2 / 1.7 \times 10^{-2}$
$= 5.9 \times 10^{-7}$ M

Now $K_H = [H_2SO_3] / P$
so $P = [H_2SO_3] / K_H$
$= 5.9 \times 10^{-7}$ M / 1.0 M atm^{-1}
$= 5.9 \times 10^{-7}$ atm

Thus, its SO_2 concentration must be 0.59 ppm.

Problem 3-15

The first reaction of relevance is:

$$CO_2(g) + H_2O(aq) \rightleftharpoons H_2CO_3(aq)$$

for which $K_H = [H_2CO_3(aq)] / P_{CO_2}$

Since $P_{CO_2} = 0.00039$ atm.

$[H_2CO_3(aq)] = K_H P_{CO_2} = 3.4 \times 10^{-2}$ mol L^{-1} atm^{-1} $\times$ 0.00039 atm
$= 1.33 \times 10^{-5}$ M.

Now, acidity results from ionization of H_2CO_3:

$$H_2CO_3 \rightleftharpoons H^+ + HCO_3^-$$

$$K_a = [H^+]\,[HCO_3^-] / [H_2CO_3]$$

We know $[H^+] = [HCO_3^-]$ for this reaction, and that the equilibrium value for $[H_2CO_3]$ $= 1.33 \times 10^{-5}$ M (see above).

Since $\dfrac{[H^+]\,[HCO_3^-]}{[H_2CO_3]} = K_a$

thus, $[H^+]^2 = K_a [H_2CO_3]$
$= 4.5 \times 10^{-7} \times 1.33 \times 10^{-5}$

so $[H^+] = 2.45 \times 10^{-6}$

and thus pH = 5.61.

Recalculating with $P_{CO_2} = 0.00056$ atm gives pH = 5.53

Problem 3-16

By analogy with chlorine, the reactions are

$$I + O_3 \longrightarrow IO + O_2$$

$$2\ IO \longrightarrow I_2O_2$$

Problem 3-17

Since ultrafine particles have diameters less than 0.1 micrometer, the correct symbol is $PM_{0.10}$.

Since the TSP index includes *all* particles, i.e., up to infinite diameter (in principle), its symbol is PM_{∞}. (Since the upper limit for suspended particles in air has been stated to be about 100 micrometers, an acceptable alternative answer is PM_{100}).

Numerically, $PM_{0.10}$ would be smaller because it includes fewer particles in its mass value.

Problem 3-18

The surface area for each of the six faces of any cube is the square of its length, so when the length is 3k,

$$\text{Total Surface Area} = 6\,(3k)^2 = 54\,k^2$$

Similarly, for each of the smaller cubes the total surface area is $6\,k^2$, and because there are 27 of them, the total surface area of all the small cubes taken together is $27 \times 6\,k^2 = 162\,k^2$.

$$\text{Thus, } \frac{\text{Surface area for the mass split into small cubes}}{\text{Surface area for mass as one cube}} = \frac{162\,k^2}{54\,k^2} = 3$$

Thus, the total surface area for a given mass becomes larger as the particles into which it is split become smaller.

Box 3-1, Problem 1

(a) We must convert the denominator from a billion to a million basis.

$$\frac{32 \text{ parts pollutant}}{1 \text{ billion parts air}} \times \frac{1 \text{ billion parts air}}{1000 \text{ million parts air}}$$

$$= \frac{0.032 \text{ parts pollutant}}{1 \text{ billion parts air}} = 0.032 \text{ ppm}$$

(b) Since the final concentration unit involves molecules, it is convenient to interpret "parts" in terms of molecules.

$$\text{i.e., } 32 \text{ ppb} = \frac{32 \text{ molecules of pollutant}}{10^9 \text{ molecules of air}}$$

Convert 10^9 molecules of air to volume of air using Avogadro's constant and the Ideal Gas Law:

$$n = 10^9 \text{ molecules} \times \frac{1 \text{ mole}}{6.02 \times 10^{23} \text{molecules}} = 1.66 \times 10^{-15} \text{ moles of air}$$

$$PV = nRT, \text{ so} \quad V = nRT/P$$
$$= 1.66 \times 10^{-15} \text{ mol} \times 0.082 \text{ L atm mol}^{-1} \text{ K}^{-1} \times 298 \text{ K}/1.0 \text{ atm}$$
$$= 4.06 \times 10^{-14} \text{ L}$$

Convert the value in L to that in cm^3

$$4.06 \times 10^{-14} \text{L} \times \frac{1000 \text{ cm}^3}{1 \text{ L}} = 4.06 \times 10^{-11} \text{ cm}^3$$

$$\text{Thus, the concentration is } \frac{32 \text{ molecules of pollutant}}{4.06 \times 10^{-11} \text{ cm}^3} = 7.9 \times 10^{11} \text{ molecules/cm}^3$$

(c) Using intermediate results from (b), we can switch to the molarity scale, i.e., moles of pollutant per liter of air:

$$32 \text{ molecules pollutant} \times \frac{1 \text{ mole pollutant}}{6.02 \times 10^{23} \text{ molecules}} = 5.3 \times 10^{-23} \text{ moles pollutant}$$

$$\text{Concentration is } \frac{5.3 \times 10^{-23} \text{ moles}}{4.06 \times 10^{-14} \text{ L}} = 1.3 \times 10^{-9} \text{ moles/L}$$

Box 3-1, Problem 2

Conversion to moles per liter requires a conversion of molecules pollutant to moles pollutant, and cm^3 of air to L of air:

$$\frac{6.0 \times 10^{14} \text{ molecules pollutant}}{1 \text{ cm}^3 \text{ air}} \times \frac{1 \text{ mole pollutant}}{6.02 \times 10^{23} \text{ molecules}} \times \frac{1 \text{ cm}^3}{0.001 \text{ L air}}$$

$$= 1.0 \times 10^{-6} \text{ moles pollutant/L air} = 1.0 \times 10^{-6} \text{ M.}$$

Conversion to the ppm scale is done most easily by interpreting ppm as molecules of pollutant/million molecules of air; thus we need only establish the number of molecules in 1 cm^3 of air.

Using the Ideal Gas Law, for 1 cm^3 = 0.001 L

Thus, $n = PV/RT = 1.00 \text{ atm} \times 0.001 \text{ L}/0.082 \text{ L atm mol}^{-1} \text{K}^{-1} \times 298 \text{ K}$
$= 4.09 \times 10^{-5} \text{ mol}$

Using Avogadro's constant to convert moles to molecules:

$$4.09 \times 10^{-5} \text{ mol} \times \frac{6.02 \times 10^{23} \text{ molecules}}{1 \text{ mol}} = 2.46 \times 10^{19} \text{ molecules}$$

Thus, the number of millions of air molecules is 2.46×10^{13}

$$\text{Concentration is } \frac{6.0 \times 10^{14} \text{ molecules pollutant}}{2.46 \times 10^{13} \text{ million molecules of air}} = 24 \text{ ppm}$$

Box 3-1, Problem 3

(a) Given the final units, it is convenient to interpret 40 ppb as 40 molecules of ozone per 10^9 molecules of air.

Convert the air data into volume in cm^3:

$$n = 10^9 \text{ molecules} \times \frac{1 \text{ mole}}{6.02 \times 10^{23} \text{ molecules}} = 1.66 \times 10^{-15} \text{ moles of air}$$

Using the Ideal Gas Law,

$$V = \frac{nRT}{P} = \frac{1.66 \times 10^{-15} \text{ mol} \times 0.082 \text{ L atm mol}^{-1} \text{K}^{-1} \times 300 \text{ K}}{0.95 \text{ atm}}$$
$$= 4.3 \times 10^{-14} \text{L}$$
$$= 4.3 \times 10^{-11} \text{cm}^3$$

Concentration is 40 molecules ozone/4.3×10^{-11} cm^3 air
$= 9.3 \times 10^{11}$ molecules ozone / cm^3 air

(b) It is convenient to use the volume information from part (a) and to convert the molecules of O_3 into its mass.

$$40 \text{ molecules } O_3 \times \frac{1 \text{ mol } O_3}{6.02 \times 10^{23} \text{ molecules } O_3} = 6.64 \times 10^{-23} \text{ mol } O_3$$

$$6.64 \times 10^{-23} \text{mol } O_3 \times \frac{48.0 \text{ g } O_3}{1 \text{ mole } O_3} = 3.19 \times 10^{-21} \text{ g } O_3$$

$$\text{The concentration of } O_3 \text{ is } \frac{3.19 \times 10^{-21} \text{g } O_3}{4.30 \times 10^{-14} \text{ L}} \times \frac{1000 \text{ L}}{1 \text{ m}^3} \times \frac{10^6 \mu\text{g } O_3}{1 \text{ g } O_3}$$
$$= 74.2 \ \mu\text{g } O_3 / \text{m}^3 \text{of air}$$

Box 3-1, Problem 4

The original concentration is:

$$\frac{1000 \times 10^{-6} \text{ grams of CO}}{1 \text{ m}^3 \text{ of air}}$$

To convert to the ppm scale, we need to know the number of molecules of CO for each 1 million molecules of air, or the volume of CO per 1 million volumes of air.

Using the volume-based definition (because the denominator of the concentration is already expressed as a volume), we first convert the mass of CO into its volume at the stated temperature and pressure:

$$n = 1000 \times 10^{-6} \text{ g CO} \times \frac{1 \text{ mole CO}}{28.0 \text{ g CO}} = 3.6 \times 10^{-5} \text{ moles CO}$$

$$V = nRT/P = \frac{3.6 \times 10^{-5} \text{ moles} \times 0.082 \text{ L atm mol}^{-1} \text{ K}^{-1} \times (273 + 17 \text{ K})}{1.04 \text{ atmospheres}}$$

$$= 8.2 \times 10^{-4} \text{ L of CO}$$

Thus because $1 \text{ m}^3 = 1000$ L of air, the concentration of CO is:

$$\frac{8.2 \times 10^{-4} \text{ L CO}}{1000 \text{ L air}} = \frac{0.82 \text{ L CO}}{10^6 \text{ L air}}$$

Thus the concentration of CO is 0.82 ppm.

To express the concentration on the molecules of CO per cm^3 of air scale, return to the mass of CO in moles and convert it to molecules using Avogadro's number:

$3.6 \times 10^{-5} \text{ mol} \times 6.02 \times 10^{23} \text{ molecules mol}^{-1} = 2.2 \times 10^{19}$ molecules

Since 1 meter = 100 centimeters, $1 \text{ m}^3 = (10^2)^3 = 10^6 \text{ cm}^3$

Thus the concentration is 2.2×10^{19} molecules / 10^6 cm^3

$= 2.2 \times 10^{13}$ molecules / cm^3

Box 3-3, Problem 1

The distribution for d^2 times that for the number distribution would presumably be similar to that for d^3 times the number, but displaced somewhat to its left because it is a lesser power of d, the x-axis variable. However it will probably be closer to the mass (d^3) distribution than to the number one (corresponding to d^0). Thus it will probably consist of two peaks, lying between the number and mass distributions and closer to the latter, at approximately 0.2 and 2 μm.

Green Chemistry Problems

1. (a) 3. The design of chemicals that are less toxic than current alternatives or inherently safer with regard to accident potential.

   (b) 1. Prevention of waste

   7. A raw material feedstock should be renewable rather than depleting whenever technically and economically practical.

2. The reduction of VOC emissions, and the use of renewable feedstocks/reduction in the use of petroleum.

3. The presence of the 1,4 C=C double bonds results in a diallylic position. A hydrogen in this diallylic position is easily removed because it results in a resonance-stabilized free radical. Yes, because linoleic acid has hydrogens that are diallylic. No, stearic acid is saturated.

4. (a) Both gasoline and kerosene are volatile hydrocarbons (VOCs) and their vapors would contribute to pollution of the troposphere. Both gasoline and kerosene are flammable liquids and their use would result in reduced worker safety.

   (b) Perc is also a VOC and would contribute to tropospheric pollution, but it is not flammable.

   (c) Carbon dioxide is a greenhouse gas, but the carbon dioxide used for dry cleaning is waste carbon dioxide and it is recycled in the dry cleaning.

5. (a) Area 2. Alternative reaction conditions for green chemistry.

   (b) Principle 1. Prevention of waste.

   Principle 4. Preserving efficacy of function while reducing toxicity.

6. For the first pair of ions, the chloride ion (the anion) has a localized charge, but the cation has a delocalized charge (about the imidazole ring) and two non-polar groups. Both of these characteristics of the cation help to weaken the interaction between the oppositely charged ions. In the second pair, the anionic $AlCl_4$ is bulky with the charge dispersed over the four chlorides (even though the formal charge is actually on the Al), while the cation has a bulky non-polar butyl group attached to the charge-bearing nitrogen. In the third pair, both the anion and the cation have large non-polar bulky groups, and in addition the charge on the anion is delocalized over both of the oxygens. In the fourth pair, the charge on the anion is delocalized over three oxygens, and the cation has four nonpolar bulky groups surrounding the positively charged nitrogen.

7. (a) Area 3. Design of safer chemicals.

   (b) Cellulose:

   7. A raw material feedstock should be renewable rather than depleting whenever technically and economically practical.

10. Chemical products should be designed so that at the end of their function they do not persist in the environment but break down into innocuous degradation products.

Microwave heating:

1. It is better to prevent waste than to treat or clean up waste after it is formed.
6. Energy requirements should be recognized for their environmental and economic impacts and should be minimized. Synthetic methods should be conducted at ambient temperature and pressure.

Ionic liquids:

5. The use of auxiliary substances (e.g., solvents, separation agents, etc.) should be made unnecessary whenever possible and innocuous when used.
12. Substances and the form of a substance used in a chemical process should chosen so as to minimize the potential for chemical accidents, including releases, explosions, and fires.

Additional Problems

1. First we should convert the ppb concentrations to molecules cm^{-3}, the concentration units applicable to the rate constant data.

$$\text{Since 40 ppb of } O_3 = \frac{40 \text{ moles of } O_3}{10^9 \text{ moles of air}}$$

we can switch moles of O_3 directly to molecules of it:

$$40 \text{ moles} \times \frac{6.02 \times 10^{23} \text{ molecules}}{1 \text{ mole}} = 2.41 \times 10^{25} \text{ molecules of } O_3$$

To convert the moles of air to volume in L, we use the ideal gas law at the stated temperature and pressure:

$$V = nRT/P = 10^9 \times 0.082 \times 300/1.0 = 2.46 \times 10^{10} \text{ L}$$

Since 1 L = 1000 cm^3, then the volume of air is 2.46×10^{13} cm^3

$$\text{Thus the concentration of ozone is } \frac{2.41 \times 10^{25} \text{ molecules}}{2.46 \times 10^{13} \text{ cm}^3} = 9.8 \times 10^{11} \text{ molecules cm}^{-3}$$

Since the ppb concentration of NO is twice that of ozone, its converted concentration will be 1.96×10^{12} molecules cm^{-3}.

Finally, we note that the ppb concentration of O_2 in air is 2.1×10^8 (because by volume it constitutes 21% of air); by proportion to the ozone conversion above, its concentration is 5.2×10^{18} molecules cm^{-3}.

For the reaction

$$NO + O_3 \longrightarrow NO_2 + O_2$$

it follows that

$$\begin{aligned} \text{Rate} &= k\,[NO]\,[O_3] \\ &= 2 \times 10^{-14} \times 1.96 \times 10^{12} \times 9.8 \times 10^{11} \\ &= 4 \times 10^{10} \text{ molecules cm}^{-3}\text{ sec}^{-1} \end{aligned}$$

For the reaction

$$2\,NO + O_2 \longrightarrow 2\,NO_2$$

it follows from the Collision theory of reaction rates that

$$\begin{aligned} \text{Rate} &= k'\,[NO]^2\,[O_2] \\ &= 2 \times 10^{-38} \times (1.96 \times 10^{12})^2 \times 5.2 \times 10^{18} \\ &= 4 \times 10^{5} \text{ molecules cm}^{-3}\text{ sec}^{-1} \end{aligned}$$

Clearly the reaction of NO with ozone is dominant over that with molecular oxygen.

2. (a) First convert the rate constant to a numerical value, using:

$$t = 30^\circ\text{C, so T} = 273 + 30 = 303\text{ K.}$$

$$\begin{aligned} k &= 5 \times 10^{-13}\, e^{-300/T} \\ &= 5 \times 10^{-13}\, e^{-300/303} \\ &= 1.9 \times 10^{-13} \end{aligned}$$

Since the units of k are given as molecule^{-1} cm^3 sec^{-1}, we must convert the CO concentration from ppm to molecules cm^{-3}. First, interpret 20 ppm as 20 L of CO in 10^6 L of air.

Convert the volume of CO to molecules of it using the Ideal Gas Law and Avogadro's constant:

$$\begin{aligned} n &= PV/RT = 1.0\text{ atm} \times 20\text{ L}/0.082\text{ L atm mol}^{-1}\text{ K}^{-1} \times 303\text{ K} \\ &= 0.80\text{ moles CO} \end{aligned}$$

$$\begin{aligned} \text{Molecules CO} &= 0.80\text{ moles} \times 6.02 \times 10^{23}\text{ molecules/mole} \\ &= 4.8 \times 10^{23}\text{ molecules CO} \end{aligned}$$

Now convert 10^6 L of air into cm^3:

$$10^6\text{ L} \times \frac{1000\text{ cm}^3}{1\text{ L}} = 10^9\text{ cm}^3$$

Thus the concentration of CO is 4.8×10^{23} molecules/10^9 cm^3, or 4.8×10^{14} molecules/cm^3. Since the reaction involves one molecule of CO and one of OH, its rate law is:

$$\begin{aligned} \text{Rate} &= k\,[CO]\,[OH] \\ &= 1.9 \times 10^{-13} \times 4.8 \times 10^{14} \times 8.7 \times 10^{6} \\ &= 8.0 \times 10^{8}\text{ molecules cm}^{-3}\text{ sec}^{-1} \end{aligned}$$

(b) Rate $= k\,[CO]\,[OH]$

but if [CO] is essentially constant , we can define the new constant

$$k' = k\,[CO]$$

$$\text{so } k' = k\,[CO] = 1.9 \times 10^{-13}\ \text{molecule}^{-1}\ \text{cm}^3\ \text{sec}^{-1} \times 4.8 \times 10^{14}\ \text{molecule cm}^{-3}$$

$$= 91\ \text{s}^{-1}$$

Now Rate $= k'\,[OH]$

Since from elementary kinetics, for first-order reactions, the half-life

$$t_{1/2} = 0.69\ /\ \text{rate constant}$$

thus $t_{1/2} = 0.69 / k' = 0.69 / 91\ \text{s}^{-1} = 0.0076$ sec

Thus the half-life of OH is 0.0076 seconds.

3. (a) From the description in the question, the reaction step is

$$O + N_2 \longrightarrow NO + N$$

From the Collision theory of reaction rates, the rate equation for the step will be first order in O and first order in N_2 (because 1 of each is involved in the step).

$$\text{Rate} = k\,[O]\,[N_2]$$

(b) We are told that

$$k = 9.7 \times 10^{10} \text{ at } 800°\text{C (or 1073 K)}$$

and that the activation energy $E = 315\ \text{kJ mol}^{-1}$.

The variation of a rate constant with temperature is given by the Arrhenius equation

$$k = Ae^{-E/RT}$$

in which, in energy units, $R = 8.3\ \text{J K}^{-1}\ \text{mol}^{-1}$.

To deduce k at a different temperature, we could first solve this equation for A from the k, E, and T values applied:

$$A = k\,e^{E/RT} = 9.7 \times 10^{10}\, e^{315{,}000/8.3\ \times\ 1073}$$
$$= 2.2 \times 10^{26}$$

Thus at $t = 1100°$ C, then $T = 1373$ and

$$k = Ae^{-E/RT}$$
$$= 2.2 \times 10^{26}\, e^{-315{,}000/8.3\ \times\ 1373}$$
$$= 2.2 \times 10^{14}$$

Thus the factor by which k increases is $2.2 \times 10^{14}/9.7 \times 10^{10} = 2.2 \times 10^{3}$.

4. From the balanced equation for the reaction, thus

$$K = [NO]^2 / [N_2] [O_2]$$

From the data in Chapter 1 we know for air:

$$[N_2] = 0.78 \text{ atm}$$
$$[O_2] = 0.21 \text{ atm}$$

∴ By rearranging the equation and substituting, we obtain:

$$[NO]^2 = K [N_2] [O_2] = K \times 0.78 \times 0.21$$
$$= 0.16\, K$$

$$\therefore [NO] = 0.40 \sqrt{K}$$

When $K = 10^{-14}$, then $\sqrt{K} = 10^{-7}$ and $[NO] = 4 \times 10^{-8}$ atm

When $K = 10^{-30}$, then $\sqrt{K} = 10^{-15}$ and $[NO] = 4 \times 10^{-16}$ atm

Since the exit [NO] > > calculated value, the reaction does not successfully remain in equilibrium as the temperature is lowered from that in the engine. The NO is "trapped" in that form.

5. (a) From the description in the statement of the problem, the unbalanced equation is:

$$O_3 + KI \longrightarrow I_2 + O_2 + KOH$$

To supply the hydrogen in the product, H_2O must also be consumed in the aqueous-phase reaction

$$O_3 + KI + H_2O \longrightarrow I_2 + O_2 + KOH$$

By inspection (or using a redox balancing procedure) the balanced equation is:

$$O_3 + 2\,KI + H_2O \longrightarrow I_2 + O_2 + 2\,KOH$$

(b) From the mass of KI used in the reaction, we can deduce the amount, in moles, of O_3 that reacted using the balanced equation

$$17.0 \times 10^{-6} \text{ g KI} \times \frac{1 \text{ mole KI}}{166.0 \text{ g KI}} \times \frac{1 \text{ mole } O_3}{2 \text{ moles KI}} = 5.1 \times 10^{-8} \text{ moles } O_3$$

The ppb concentration of ozone can be obtained by obtaining the number of moles of air in the 10.0 L sample using the Ideal Gas Law:

$$n = PV/RT = 1.0 \times 10/0.082 \times 300 = 0.41 \text{ moles}$$

Thus the ppb concentration of ozone is:

$$\frac{5.1 \times 10^{-8} \text{ moles } O_3}{0.41 \times 10^{-9} \text{ billion moles air}} = 1.2 \times 10^2$$

6. The three-step mechanism is

$$NO_2 \longrightarrow NO + O \quad \text{..... (1)}$$

$$O + O_2 \longrightarrow O_3 \quad \text{..... (2)}$$

$$NO + O_3 \longrightarrow NO_2 + O_2 \quad \text{..... (3)}$$

Steady-state for ozone: $d[O_3] / dt = 0 = k_2 [O][O_2] - k_3 [NO][O_3]$

Steady-state for O: $d[O] / dt = 0 = k_1 [NO_2] - k_2 [O][O_2]$

If we add the right-hand sides of the two 0= equations, the common term $k_2 [O][O_2]$ cancels, and we have

$$-k_3 [NO][O_3] + k_1 [NO_2] = 0$$

which after rearrangement gives us $[NO_2] / [NO] = k_3 [O_3] / k_1$

7. The sulfur is converted from S to SO_2 to H_2SO_4, so the moles of sulfur equal that of H_2SO_4. The moles of H^+ in the titration equals the molarity times volume (in L) of the NaOH:

$$\text{Moles NaOH} = MV = 0.114 \frac{\text{moles}}{\text{L}} \times 0.0441 \text{ L} = 0.00503 \text{ moles}$$

The balanced reaction between NaOH and H_2SO_4 is

$$2\,NaOH + H_2SO_4 \longrightarrow 2\,H_2O + Na_2SO_4$$

Thus converting to moles H_2SO_4, then to moles S, then to mass S we have:

$$0.00503 \text{ moles NaOH} \times \frac{1 \text{ mole } H_2SO_4}{2 \text{ moles NaOH}} \times \frac{1 \text{ mole S}}{1 \text{ mole } H_2SO_4} \times \frac{32.1 \text{ g S}}{1 \text{ mole S}} = 0.0807 \text{ g S}$$

Thus the percent sulfur is $(0.0807/8.05) \times 100\% = 1.0\%$

8. a) $H_2SO_3 + SO_3^{2-} \longrightarrow 2\,HSO_3^-$

b) Bubble the gas containing SO_2 through an aqueous solution containing sulfite ion to capture it in solution as HSO_3^-. Later, when the solution is concentrated, heat it and/or add concentrated strong acid to it to reverse the reaction, and decompose the acid, and release the SO_2 as a concentrated stream.

9. Since the SO_2 concentration in air is 2.0 ppb, its partial pressure there is 2.0×10^{-9} atm. The gas-solution equilibrium of interest is

$$SO_2\ (g) + H_2O\ (aq) \rightleftharpoons H_2SO_3\ (aq)$$

for which the Henry's Law constant $K_H = [H_2SO_3\ (aq)] / P_{SO_2}$

We can solve this latter equation for the equilibrium $[H_2SO_3\ (aq)]$:

$$[H_2SO_3\ (aq)] = K_H\ P_{SO_2} = 1.0\ M\ atm^{-1} \times 2.0 \times 10^{-9}\ atm = 2.0 \times 10^{-9}\ M$$

The weak acid equilibrium is

$$H_2SO_3 \rightleftharpoons H^+ + HSO_3^-$$

for which $K_a = [H^+]\,[HSO_3^-]\ /\ [H_2SO_3\ (aq)]$

Since we know equilibrium values here for $[H_2SO_3\ (aq)]$ and $[H^+]$, we can solve for $[HSO_3^-] = K_a\ [H_2SO_3\ (aq)]\ /\ [H^+]$

Substituting the values for the two terms in the numerator,

we have

$$[HSO_3^-] = 1.7 \times 10^{-2} \times 2.0 \times 10^{-9}\ /\ [H^+] = 5.8 \times 10^{-13} / [H^+]$$

If the pH = 4, then $[H^+] = 1.0 \times 10^{-4}$, so by substitution, $[HSO_3^-] = 5.8 \times 10^{-9}$

The total solubility of SO_2 equals the equilibrium concentrations of H_2SO_3 (aq) and HSO_3^-, which in this case equals $2.0 \times 10^{-9} + 5.8 \times 10^{-9} = 7.8 \times 10^{-9}$ M.

Similarly, when the pH = 5, the solubility is 6.0×10^{-8} M, and when pH = 6, the solubility is 5.8×10^{-7} M.

10. (a) If the pH falls, the concentration of H^+ increases, which by LeChatelier's Principle drives the equilibrium to the left, decreasing the concentration of HSO_3^- in the reaction

$$H_2SO_3 \rightleftharpoons H^+ + HSO_3^-$$

Since the concentration of HSO_3^- decreases, its rate of oxidation by ozone will decrease.

(b) As in part (a), decreasing the pH decreases the concentration of HSO_3^-.

However, increasing $[H^+]$ drives the equilibrium in the reaction below to the right, producing more of the ion that oxidizes HSO_3^-.

$$H_2O_2 + H^+ \rightleftharpoons H_3O_2^+$$

The two opposing effects on the rate cancel, so pH should have little effect in the rate of oxidation by hydrogen peroxide.

CHAPTER 4

The Environmental and Health Consequences of Polluted Air—Outdoors and Indoors

Problem 4-1

The unbalanced half-reaction is:

$$NH_4^+ \longrightarrow NO_3^-$$

We can assume that the reaction occurs in acid, because we are told that H^+ is produced. Any standard redox balancing scheme can be used to balance this reaction. For example, we could first balance the oxygen atoms by adding H_2O to the left side.

$$NH_4^+ + 3\,H_2O \longrightarrow NO_3^-$$

Then balance the H's by supplying H^+ to the right side.

$$NH_4^+ + 3\,H_2O \longrightarrow NO_3^- + 10\,H^+$$

Finally, balance the charge by adding electrons to the right:

$$NH_4^+ + 3\,H_2O \longrightarrow NO_3^- + 10\,H^+ + 8e^-$$

Thus we see that H^+ is indeed produced in the reaction.

Problem 4-2

Since the formula for sulfate ion is SO_4^{2-}, the unbalanced half-reaction is

$$SO_2 \longrightarrow SO_4^{2-}$$

The oxidation number of the sulfur in SO_2 must be +4 because each O atom is –2, and that in SO_4^{2-} must be +6 because each O is –2, giving –8 and the net charge is –2. Thus the half-reaction is a 2-electron loss process, converting +4 sulfur to +6:

$$SO_2 \longrightarrow SO_4^{2-} + 2\,e^-$$

To balance the –4 charge on the right side with zero on the left, we add 4 H^+ ions to the right:

$$SO_2 \longrightarrow SO_4^{2-} + 2\,e^- + 4\,H^+$$

To balance O atoms, we note that the left side is deficient in 2 O, so we add 2 H_2O molecules to that side:

$$SO_2 + 2\,H_2O \longrightarrow SO_4^{2-} + 2\,e^- + 4\,H^+$$

The half-reaction is now balanced in atoms and charge.

Problem 4-3

For spherical particles, the surface area is $4\pi r^2 = \pi d^2$ (because $2r = d$). Thus, for each particle of diameter 10 μm, the surface area is 100π μm^2, whereas for each particle of diameter 0.1 μm, the surface area is 0.01π μm^2.

We must now calculate the relative numbers of particles of the two types. Assuming that the particles have equal densities, then

$$\text{Volume} \propto \text{mass}$$

$$\text{but volume} = 4\pi r^3 / 3 = \pi d^3 / 6$$

Thus, the relative volumes of each particle of 10 μm and 0.1 μm are

$$\frac{\text{Volume of 10 μm particles}}{\text{Volume of 0.1 μm particles}} = \frac{(10\ \mu m)^3}{(0.1\ \mu m)^3} = 10^6$$

Thus, the mass of each large particle is 1 million times that of each smaller one. Since, however, the total mass of the larger particles is only $95 / 5 = 19$ times that of the smaller ones, the ratio of the *number* of small particles to large ones is $10^6 / 19 = 5.3 \times 10^4$ to 1. Since the ratio of surface area of a large to a small particle is $(100\pi / 0.01\pi) = 10^4$, the ratio of the total surface area of the small to the large particles is $5.3 \times 10^4 / 10^4 = 5.3$ to 1. Thus, eliminating all large particles reduces the *total* surface area of all particles by $1.0 / 6.3 = 0.16$, i.e., by only 16%.

Additional Problems

1. The acidity in the sample is presumably due to a combination of both H_2SO_4 and HNO_3, both of which are fully ionized under these pH conditions. Thus

$$[H^+] = 2\,[H_2SO_4] + [HNO_3]$$

(where the acid concentrations refer to values before ionization occurs).

Since pH $= 4.2$, $[H^+] = 10^{-4.2} = 6.3 \times 10^{-5}$ M

Since the sulfur concentration is 0.000010 M $= 1.0 \times 10^{-5}$ M,

thus $[HNO_3] = 6.3 \times 10^{-5} - 2 \times 1.0 \times 10^{-5}$
$= 4.3 \times 10^{-5}$ M

Since the ratio of nitric to all acid is $(4.3 \times 10^{-5}/6.3 \times 10^{-5}) = 0.68$, the sample probably originated in western North America where the nitrogen acid dominates acid rain.

2. Since pH $= 4.0$, thus $[H^+] = 1.00 \times 10^{-4}$ M. Since half the acidity is due to nitric acid, thus $[HNO_3] = 0.50 \times 10^{-4}$ M.

   Half the acidity is due to H_2SO_4, which contributes two H^+ for each H_2SO_4, so

   $[H_2SO_4] = 0.25 \times 10^{-4}$ M

   Since 1 mole of HNO_3 results from 1 mole of NO, for 1 L of rain we need 0.50×10^{-4} moles of NO:

$$0.50 \times 10^{-4} \text{moles NO} \times \frac{30.0 \text{ g NO}}{1 \text{ mole NO}} = 0.0015 \text{ g NO}$$

   Since 1 mole of H_2SO_4 results from 1 mole of SO_2, for 1 L of rain we need 0.25×10^{-4} moles of SO_2:

$$0.25 \times 10^{-4} \text{moles } SO_2 \times \frac{64.1 \text{ g } SO_2}{1 \text{ mole } SO_2} = 0.0016 \text{ g } SO_2$$

3. The volume of the lake in cubic meters is:

$$3000 \text{ m} \times 8000 \text{ m} \times 100 \text{ m} = 2.4 \times 10^9 \text{ m}^3$$

   Since 1 m^3 contains 1000 L, the lake's volume is 2.4×10^{12} L.

   Next, convert pHs to H^+ concentrations and deduce the moles of H^+ present under the two sets of conditions.

   Since $-\log [H^+] = 4.5$, thus $[H^+] = 3.16 \times 10^{-5}$ M.

   Similarly, pH $= 6.0$ corresponds to $[H^+] = 1.0 \times 10^{-6}$ M.

   Thus the *change* in H^+ in each liter (as a result of the reaction with $CaCO_3$) must be the difference between these two values, i.e., 3.06×10^{-5} moles. Multiplying this value by the lake's volume gives the number of moles of H^+ that must be reacted.

$$3.06 \times 10^{-5} \frac{\text{moles}}{1\text{ L}} \times 2.4 \times 10^{12}\text{ L} = 7.3 \times 10^{7}\text{ moles H}^{+}$$

The reaction of H^+ with $CaCO_3$ to yield CO_2 is:

$$2\,H^+ + CaCO_3 \longrightarrow Ca^{2+} + H_2O + CO_2$$

Thus the moles of $CaCO_3$ required is half that of H^+. Since the molar mass of $CaCO_3$ is 100.1 grams, the mass of it that must be added to the lake is:

$$0.5 \times 7.3 \times 10^{7}\text{ moles} \times \frac{100.1\text{ grams}}{1\text{ mole}} = 3.7 \times 10^{9}\text{ grams}$$

4. The reaction of relevance is

$$HSO_4^- \rightleftharpoons H^+ + SO_4^{2-}$$

$$\text{so } K_a(HSO_4^-) = \frac{[H^+][SO_4^{2-}]}{[HSO_4^-]}$$

where all concentrations are equilibrium values. At a given $[H^+]$, the fraction of HSO_4^- ionized is

$$\text{Fraction} = \frac{[SO_4^{2-}]}{[HSO_4^-]_0} = \frac{[SO_4^{2-}]}{[HSO_4^-] + [SO_4^{2-}]}$$

Solving for $[HSO_4^-]$ from the K_a equation,

thus, $[HSO_4^-] = [H^+][SO_4^{2-}] / K_a$

Substituting into the equation for fraction ionized, then

$$\text{Fraction} = \frac{[SO_4^{2-}]}{([H^+][SO_4^{2-}]/K_a) + [SO_4^{2-}]}$$

$$= \frac{1}{([H^+]/K_a) + 1}$$

Since $K_a = 1.2 \times 10^{-2}$, and since at pH = 4, $[H^+] = 10^{-4}$

thus, fraction $= 1 / (10^{-4}/1.2 \times 10^{-2} + 1) = 0.992$, i.e., 99.2%

Similarly at pH = 3, fraction $= 1 / (10^{-3} / 1.2 \times 10^{-2} + 1) = 0.923$, i.e., 92.3%. Yes, it *is* consistent with LeChatelier's Principle, because increasing $[H^+]$ pushes the equilibrium to the left, ionizing less HSO_4^-.

5. First convert the volume of inhaled air to cubic meters and then obtain the mass of particles:

$$350 \text{ L} \times \frac{1 \text{ m}^3}{1000 \text{ L}} \times \frac{10 \times 10^{-6} \text{ g}}{1 \text{ m}^3} = 3.5 \times 10^{-6} \text{ grams.}$$

From the mass of particles, we can deduce their volume because their density is supplied:

$$3.5 \times 10^{-6} \text{g} \times \frac{1 \text{ mL}}{0.5 \text{ g}} \times \frac{1 \text{ L}}{1000 \text{ mL}} \times \frac{1 \text{ m}^3}{1000 \text{ L}} = 7.0 \times 10^{-12} \text{ m}^3$$

The surface area S of a spherical particle is $4 \pi r^2$ and its volume V is $(4/3) \pi r^3$, so the ratio of surface area to volume for each particle and the sample as a whole is:

$$\frac{S}{V} = \frac{4 \pi r^2}{(4/3) \pi r^3} = \frac{3}{r}$$

$$\text{Thus } S = 3 V/r = 3 \times 7 \times 10^{-12} \text{ m}^3/0.5 \times 10^{-6} \text{ m}$$
$$= 4.2 \times 10^{-5} \text{ m}^2$$

Finally, we multiply this hourly surface area by the number of hours in a year to obtain the annual load.

$$\frac{4.2 \times 10^{-5} \text{m}^2}{1 \text{ h}} \times \frac{24 \text{ h}}{1 \text{ d}} \times \frac{365 \text{ d}}{1 \text{ y}} = 0.37 \text{ m}^2/\text{y}$$

6. We need to *either* convert 100 ppb to $\mu g\ m^{-3}$, *or* convert 250 $\mu g\ m^{-3}$ to ppb. Choosing the former:

   100 ppb formaldehyde (H_2CO) = 100 mol H_2CO / 10^9 mol air

   100 mol $H_2CO \times 30.0$ g / mol $\times 10^6$ μg / g = 3.0×10^9 μg

   V of 10^9 mol air: $V = n R T / P = 10^9 \text{ mol} \times 0.082 \text{ L atm K}^{-1} \text{ mol}^{-1} \times 296 \text{ K} / 1 \text{ atm} =$ $2.43 \times 10^{10} \text{ L} \times 1 \text{ m}^3 / 1000 \text{ L} = 2.43 \times 10^7 \text{ m}^3$ of air

Therefore, the detection threshold of 100 ppb H_2CO corresponds to 3.0×10^9 μg / 2.43×10^7 m^3 = 123 $\mu g\ m^{-3}$. Thus a concentration of 250 $\mu g\ m^{-3}$ is well above the detection threshold, and should be detectable by a typical human.

7. First deduce the volume of the room in cubic meters and then in liters:

$$V = 4 \text{ m} \times 5 \text{ m} \times 2 \text{ m} = 40 \text{ m}^3 \times \frac{1000 \text{ L}}{1 \text{ m}^3} = 40{,}000 \text{ L}$$

Since the concentration of formaldehyde is to be 0.50 ppm, there are 0.50 liters of it per one million liters of air; thus the volume of pure formaldehyde is given by:

$$40{,}000 \text{ L air} \times \frac{0.50 \text{ L formaldehyde}}{10^6 \text{ L air}} = 0.020 \text{ L}$$

Using the ideal gas law and the usual pressure and temperature, we can calculate the moles of formaldehyde

$$n = PV/RT = 1.0 \times 0.02/0.082 \times 300 = 0.00081 \text{ moles}$$

Since the molar mass of formaldehyde, H_2CO is 30.0, the mass of formaldehyde is:

$$0.00081 \text{ moles} \times \frac{30.0 \text{ g}}{1 \text{ mole}} = 0.024 \text{ grams}$$

CHAPTER 5

The Greenhouse Effect

Problem 5-1

Since the rate of energy release is proportional to the fourth power of the Kelvin temperature, and since 0°C = 273 K and 17°C = 290 K, the ratio of the rates of release is $(273 / 290)^4$, the value of which is 0.785.

The Kelvin temperature T at which the rate of release is twice that at 0°C would be the solution to the equation

$$(T / 273)^4 = 2.0$$

Raising both sides of the equation to the power of 1/4 gives

$$T / 273 = 1.189$$

So T = 325 K, which is equivalent to 52°C.

Problem 5-2

H_2 and Cl_2 will *not* absorb IR light because there cannot be a change in the dipole moment during the vibration; the dipole moment must always be zero for such symmetrical molecules. All the others will absorb IR light because, during at least some of their vibrations, the dipole moment will change. Note that (d), ozone, has a bent structure in which the terminal oxygen atoms are not equivalent to the central one, so the oxygen-oxygen bonds can be polar and the molecule has a dipole moment.

Problem 5-3

The diatomic molecules that could absorb IR light are CO and NO (see Problem 5-2). Since they do not absorb much light in the thermal IR region, this implies that their stretching frequencies must lie outside the thermal IR region.

Problem 5-4

Using the formula relating energy to wavelength in Chapter 1, that is

$$E = 119627/\lambda$$

will produce E in kJ mol^{-1} if λ is expressed in nm. Since 1 micrometer = 1000 nm

$$\text{it follows that } 15.0\ \mu m = 15{,}000\ nm$$
$$4.26\ \mu m = 4260\ nm$$

$$\therefore E = 119627/15000 = 8.0\ kJ\ mol^{-1} = 8000\ J\ mol^{-1}$$
$$E = 119627/4260 = 28.1\ kJ\ mol^{-1} = 28{,}100\ J\ mol^{-1}$$

Since each mole contains 6.02×10^{23} molecules, the energies per molecule are:

$$8000\ J\ mol^{-1} \times \frac{1\ mole}{6.02 \times 10^{23}\ molecules} = 1.3 \times 10^{-20}\ J\ molecule^{-1}$$

$$28{,}100\ J\ mol^{-1} \times \frac{1\ mole}{6.02 \times 10^{23}\ molecules} = 4.67 \times 10^{-20}\ J\ molecule^{-1}$$

For the reaction

$$CO_2 \longrightarrow CO + O$$

$$\text{thus } \Delta H = \Delta H_f(CO) + \Delta H_f(O) - \Delta H_f(CO_2)$$
$$= -110.5 + 249.2 - (-393.5)$$
$$= +532.2\ kJ\ mol^{-1}$$

Thus the 15 μm light supplies only

$$\frac{8.0}{532.2} = 0.015 \text{ (i.e., 1.5\%)}$$

of the energy required to dissociate CO_2, whereas the 4.26 μm light supplies

$$\frac{28.1}{532.2} = 0.053 \text{ (i.e., 5.3\%)}$$

of the energy required.

Problem 5-5

From the balanced equation it follows that 1 mole of CO_2 is released per 1 mole of $CaCO_3$ heated. From molar masses of 44.01 and 100.09 g for CO_2 and $CaCO_3$ respectively, it follows that 0.440 g of CO_2 is released per gram of $CaCO_3$ heated; thus 0.440 tonnes of CO_2 are released per tonne of $CaCO_3$ heated.

Since 1 mole of C is contained in 1 mole of CO_2, and because their molar masses are 12.01 and 44.01 respectively, then for each 44.01 grams of carbon dioxide entering the atmosphere, 12.01 g of carbon enter it; by proportion, then for each gram of CO_2, the mass of carbon that enters is (12.01/44.01) = 0.27 grams.

Problem 5-6

(a) The ppm scale gives us concentrations on a moles/moles or a volume/volume basis. Because most data in the problem are masses, presumably the moles definition is the most appropriate to use here. So first we convert all masses to moles:

First convert mass of carbon into moles of it:

$$4.7 \text{ Gt} = 4.9 \times 10^9 \text{ tonnes} \times 1000 \text{ kg / tonne} \times 1000 \text{ g / kg} = 4.7 \times 10^{15} \text{ g of C}$$

Since carbon's molar mass is 12.01, moles C = 4.7×10^{15} g C / 12.01 g C mole^{-1} = 3.9×10^{14} moles of C, which is also the change in the number of moles of CO_2.

Since air's molar mass is 29.0 and its total atmospheric mass is 5.1×10^{21} grams, we can compute the number of moles of air in the atmosphere:

$$5.1 \times 10^{21} \text{ g / } 29.0 \text{ g mole}^{-1} = 1.76 \times 10^{20} \text{ moles air}$$

The increase in ppm is the ratio of the increase in number of moles of CO_2 to the total number of millions of moles of air:

$$3.9 \times 10^{14} \text{ moles } CO_2 \text{ / } 1.76 \times 10^{14} \text{ million moles of air} = 2.2 \text{ ppm}$$

Thus the annual increase in CO_2 concentration is about 2.2 ppm.

(b) To obtain the total mass of carbon, we first use the CO_2 concentration to find moles of it:

390 ppm means 390 moles CO_2 / 1 million moles of air × 1.76×10^{14} moles of air

$$= 6.86 \times 10^{16} \text{ moles } CO_2$$

Since this is also the number of moles of carbon, we can use the molar mass of the element to determine its total atmospheric mass:

$$6.86 \times 10^{16} \text{ moles } CO_2 \times 12.01 \text{ g C / 1 mole C} = 8.24 \times 10^{17} \text{ g C}$$

Since 1 Gt = 10^{15} grams, the total carbon content is 824 Gt, in good agreement with the value of ~800 Gt in Figure 5-9.

Problem 5-7

Since the reaction is:

$$CH_4 + OH \longrightarrow CH_3 + H_2O$$

it follows that its rate law is:

$$\text{rate} = k\,[CH_4]\,[OH]$$

Substituting for k and [OH] from data in the problem

$$\text{rate} = 3.6 \times 10^{-15}\ \text{cm}^3\ \text{molecule}^{-1}\ \text{s}^{-1} \times 8.7 \times 10^5\ \text{molecule cm}^{-3}\ [CH_4]$$
$$= 3.13 \times 10^{-9}\ [CH_4]\ \text{in units of s}^{-1}$$

Substituting for the current methane concentration, then

$$\text{rate} = 3.13 \times 10^{-9}\ \text{s}^{-1} \times 1.80\ \text{ppm}$$
$$= 5.63 \times 10^{-9}\ \text{ppm/second}$$

Since 1 ppm CH_4 = 1 mole CH_4 / 10^6 moles air
= 16.05 g CH_4 / 10^6 moles air

thus, rate = $5.63 \times 10^{-9} \times 16.05$ g CH_4 sec^{-1} / 10^6 moles air
= 9.04×10^{-14} g CH_4 / sec / mole air

We now need to multiply this rate by the number of seconds in a year

$$\left[\frac{60\ \text{sec}}{1\ \text{min}} \times \frac{60\ \text{min}}{1\ \text{hr}} \times \frac{24\ \text{hr}}{1\ \text{day}} \times \frac{365\ \text{days}}{1\ \text{year}} = 3.15 \times 10^7\ \text{sec/yr}\right]$$

and the number of moles of air in the atmosphere to obtain the yearly mass of methane destroyed.
Now from data, mass of atmosphere = 5.1×10^{21} g

Average molar mass of air = 29.0 g / mole

so moles of air in atmosphere = 5.1×10^{21} g / 29.0 g mole^{-1}
= 1.76×10^{20} moles

$$\text{Thus, } CH_4 \text{ destroyed} = \frac{9.04 \times 10^{-14}\ \text{g of } CH_4}{1\ \text{sec} \times 1\ \text{moles air}} \times \frac{3.15 \times 10^7\ \text{sec}}{1\ \text{year}} \times 1.76 \times 10^{20}\ \text{moles air}$$

$$= 5.01 \times 10^{14}\ \text{g of } CH_4$$
$$= 5.01\ \text{Tg of } CH_4$$

since 1 Tg = 10^{12} grams.

Problem 5-8

The formula for the hydrate is $CH_4 \cdot 6H_2O$. We can deduce the proportion of CH_4 in one mole of the compound by comparing the molar mass of CH_4 to that of the whole compound:

$$\text{fraction of mass that is } CH_4 = \frac{\text{molar mass of } CH_4}{\text{molar mass of } CH_4 \cdot 6H_2O}$$
$$= 16.05\text{ g}/124.17\text{ g}$$
$$= 0.1293$$

Thus, in 1000 g of the hydrate, the mass of methane is

$$0.1293 \times 1000\text{ g} = 129\text{ g}$$

Problem 5-9

No, they will have no sink because they possess no C—H or multiple bonds that the OH radical can attack—and start their tropospheric oxidation—nor are they likely to be soluble in water, or to absorb visible or UV-A light because they contain no multiple bonds. They would act as greenhouse gases because C—F stretch and FCF bending vibrations fall in the thermal IR region.

For CH_3F and C_2H_5F, attack by OH would begin their oxidation and clean them out from the air in a finite time—though while still present, perhaps for decades, they would act as greenhouse gases.

Problem 5-10

The residence time T, concentration C, and rate R of concentration increase are related by the formula

$$T = C / R$$

The total amount A of the gas is proportional to C by the same factor that the rate of addition M in mass units is proportional to the rate of increase of concentration, so we can rewrite the equation as

$$T = A / M$$

Solving for A we obtain $A = T\,M = 50\text{ y} \times 2.0\text{ kg / y} = 1.0 \times 10^8\text{ kg}$

Thus the total amount of it in the atmosphere is 1×10^8 kg.

Problem 5-11

From the concentration C and the residence time T, the rate R of its addition can be calculated:

$$R = C / T = 7.0\ \mu\text{g g}_{air}^{-1} / 14\text{ y}$$
$$= 0.5\ \mu\text{g g}_{air}^{-1}\text{ y}^{-1}$$

If we multiply this result by the *total* mass of the atmosphere, the rate of addition is converted from a per-gram-of-air basis to the atmosphere as a whole:

$$0.5\ \mu g\ g_{air}^{-1}\ y^{-1} \times 5.1 \times 10^{21}\ g_{air} = 2.5 \times 10^{21}\ \mu g\ y^{-1}$$
$$= 2.5 \times 10^{15}\ g\ y^{-1}$$

We had no need to use the molar mass data for the gas or for air.

Additional Problems

1. (a) This problem is rather subtle. In particular, since SO_2 and NO_2 are *nonlinear* molecules, they each have a dipole moment, the magnitudes of which would change even during the symmetric stretch vibration. Thus, all three vibrational modes—symmetric and antisymmetric stretch and the bond angle bending—are IR active.

   (b) The SO_2 symmetric stretch would probably be the most important, since its wavelength is close to the "window" region where few other absorptions are active.

   (c) Both gases have short atmospheric lifetime (~ days), so their concentrations will not build up by accumulation, and without this feature their ability to cause much global warming is slight.

2. (a) If it had the linear NON structure, then like CO_2, the symmetric stretch vibration would be unable to absorb IR. In the NNO structure, there is a dipole moment to the molecule and it can change during the symmetrical NNO stretch; thus all 3 vibrations will absorb IR. Thus, because we know NNO has 3 IR-absorbing vibrations, its structure must be NNO and not NON.

   (b) No, because the zero dipole moment associated with the tetrahedral structure does not change during such a vibration.

3. From the increase in the CO_2 concentration, we can calculate the total amount of the gas that remained in the air. Recall that ppm can be interpreted on a moles of CO_2 per million moles of air basis; thus the increase in concentration is equivalent to 11.1 moles of CO_2 for each million moles of air. The total number of moles of air in the atmosphere can be calculated from the atmospheric mass divided by its average molar mass:

   $$\text{Moles of air in atmosphere} = 5.1 \times 10^{21}\ g / 29.0\ g / mole = 1.76 \times 10^{20}\ moles$$

   Thus the total increase in CO_2 in air is

   $$(11.1\ moles\ CO_2 / 10^6\ moles\ air) \times 1.76 \times 10^{20}\ moles\ air = 1.95 \times 10^{15}\ moles\ CO_2$$

   Since the molar mass of CO_2 equals 44.01 g, the increase in the mass of CO_2 in the air is 1.95×10^{15} moles $CO_2 \times 44.01$ g / mole $CO_2 = 8.6 \times 10^{16}$ g.

   Now the total amount of CO_2 emitted into the air was 178 Gt = 178×10^{15} g,

   so the fraction of CO_2 that remained in the air was 8.6×10^{16} g / 178×10^{15} g = 0.48, i.e., 48%.

4. If methane was growing at an annual rate of 0.6%, then because the total is 5000 Tg, the rate of input must have exceeded that of output by 0.006×5000 Tg = 30 Tg/year. Thus the rate of input must slow by 30 Tg/year to stabilize it. Since its loss rate was about 530 Tg/year, the production rate must have been 530 + 30 = 560 Tg/year, of which two-thirds or 373 Tg/year was anthropogenic in origin. Thus the percent cutback required in anthropogenic methane is:

$$(30/373) \times 100\% = 8\%$$

5. This problem can be solved numerically by evaluating F for 0.001 and double that value, and for 2 and its double, and then analyzing how F is changed in the two cases.

Since $\log_e (1 - F) = -Kcd$

thus for $Kcd = 0.001$, $\log_e (1 - F) = -0.001$

so $(1 - F) = e^{-0.001}$

$(1 - F) = 0.9990$

so $F = 0.0010$

Similarly for $Kcd = 0.002$, $(1 - F) = 0.9980$

so $F = 0.0020$

Thus for Kcd near zero, F doubles with a doubling of the concentration C.

For $Kcd = 2$, it follows $F = 0.865$
and for $Kcd = 4$, $F = 0.982$

Thus doubling Kcd in this region does not nearly double F.

6. Since $t_1 = 15°C$, thus $T_1 = 273 + 15 = 288$ K, and similarly $T_2 = 291$ K.

Since the units of K involve joules, the correct units for ΔH are also in joules:

$$\Delta H = 44{,}000 \text{ J mole}^{-1}$$

Thus $\ln (P_2/P_1) = (-44{,}000/8.3)(1/291 - 1/288) = 0.190$

and hence $(P_2/P_1) = 1.21$

Thus the vapor pressure of water increases by 21% if the liquid's temperature is raised from 18°C to 21°C. However, if the average air/surface temperature is raised to 18°C, the IR absorbed by water may decrease by less than 21% because absorption is linearly related to concentration only at very low concentrations, and because air is not saturated 100% by water vapor and the relative humidity at 21°C may be less than that at 18°C.

7. Although CO_2 emissions would decrease, the lifetime of much of the previous emissions is so long that the overall level of CO_2 would not be much affected, and so the heating by it would remain about the same. However, the sulfate aerosol lifetime is very short, so its concentration

in air would be greatly reduced by a sharp reduction in SO_2 emissions, thereby canceling its cooling effect. Thus the immediate effect on climate would probably be to significantly *increase* air temperature!

8. First, realize the meaning of 1.8 ppm in terms of moles:

   1.8 moles of CH_4 is present in 10^6 moles of air

   To determine the mass of the methane, convert its molar quantity into grams via its molar mass:

   $$1.8 \text{ mole } CH_4 \times 16.0 \text{ g } CH_4 / \text{mol } CH_4 = 28.8 \text{ g } CH_4$$

   Next, determine the number of moles of air in the atmosphere to find the factor by which the amount in 1 million moles of air should be multiplied:

   $$5.1 \times 10^{18} \text{ kg} \times 1000\text{g / kg / } 29.0 \text{ g mole}^{-1} = 1.76 \times 10^{20} \text{ moles air}$$

   Thus mass CH_4 = 28.8 g CH_4 / 1 million moles air × 1.76×10^{20} moles air = 5.1×10^{15} grams

9. Since 10 sphere diameters cover 10 D, each diameter is D and each radius r therefore is D/2. The volume of each sphere is then

   $$V = 4\pi r^3 / 3 = 4\pi (D/2)^3 / 3 = \pi D^3 / 6$$

   Since there are 100 spheres (10 rows of 10 each), the total volume is $16.7\, \pi D^3$.

   Viewed from above the spheres, the area covered by each is a circle of radius D/2. Since for each circle, area $A = \pi r^2$, then $A = \pi D^2 / 4$

   For 100 circles, the area is $100\, \pi D^2 / 4 = 25\, \pi D^2$

CHAPTER 6

Energy Use, Fossil Fuels, CO_2 Emissions, and Global Climate Change

Problem 6-1

a) If the energy growth is assumed to be exponential, then

$$E = E_0 \, e^{kt}$$

where E_0 is the value of E at time $t = 0$ and k is the fractional annual increase.

Thus $E / E_0 = e^{kt}$

But $k = 0.012$ and $t = (2035 - 2008) = 27$

So $E / E_0 = e^{0.012 \times 27} = e^{0.324} = 1.38$

Thus the total growth by 2035 would be 38%.

b) If $E / E_0 = 1.20$ (corresponding to 20% growth), then we have

$$1.20 = e^{27k}$$

Taking the ln of both sides, then $\ln (1.20) = 27k$

Thus $27k = 0.182$, so $k = 0.0068$

The annual average growth rate would have to be kept to 0.68%.

Problem 6-2

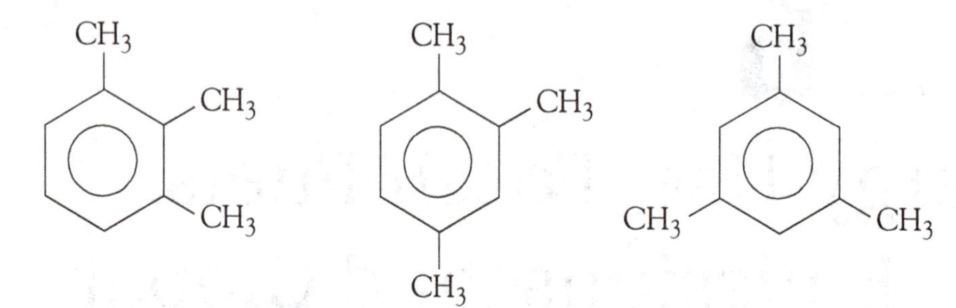

Problem 6-3

First compute the mass in grams of the CO_2, and then using the density, convert it to volume in cm^3:

$$10^{12} \text{ tonnes} \times 1000 \text{ kg / tonne} \times 1000 \text{ g / kg} \times 1 \text{ cm}^3 \text{ / g} = 1 \times 10^{18} \text{ cm}^3$$

Now convert the volume in cm^3 to km^3:

$$1 \times 10^{18} \text{ cm}^3 \times (0.01 \text{ m / cm})^3 \times (0.001 \text{ km / m})^3 = 1000 \text{ km}^3$$

The dimension of the cube is the cube root of its volume, so the length of each side is $(1000 \text{ km}^3)^{1/3}$ = 10 km.

Problem 6-4

In the reaction, one mole of CO_2 requires one mole of $CaCO_3$. Since the molar masses of CO_2 and of $CaCO_3$ are 44.01 and 100.09 grams respectively, then the ratio of the masses of $CaCO_3$ to CO_2 that react is 100.09 / 44.01 = 2.274. Thus the mass of $CaCO_3$ that will react with 1000 kg of CO_2 is 2.274 × 1000 kg = 2274 kg, i.e., 2.274 tonnes.

Problem 6-5

The balanced equations for the three reactions are shown in the text. Since

$$\Delta H = \text{Sum } \Delta H_f \text{ values for products} - \text{Sum } \Delta H_f \text{ values for reactants}$$

then for methane

$$\Delta H = \Delta H_f(CO_2) + 2\,\Delta H_f(H_2O) - \Delta H_f(CH_4) - 2\,\Delta H_f(O_2)$$

By substitution of the values given, and recalling the $\Delta H_f = 0$ for O_2, thus

$$\Delta H = -393.5 + 2 \times (-285.8) - (-74.9) - 2 \times (0) = -890.2 \text{ kJ mol}^{-1}$$

Thus because 1 mole of CO_2 is produced in this process, the moles of CO_2 per kilojoule of heat is 0.00112.

Similarly, the ΔH and moles CO_2/kJ values for CH_2 and carbon are –658.7 and 0.00152, and –393.5 and 0.00254 respectively.

Problem 6-6

Since the oxidation number of hydrogen is +1, that for carbon in C, CH, CH_2, and CH_4 must be 0, –1, –4, and –8 respectively for the sum of oxidation numbers to be zero in all cases. Since the oxidation number of oxygen in molecules is –2, that of carbon in CO_2 must be +4. Thus the changes in oxidation number in going to CO_2 must be:

C gain of 4

CH gain of 5

CH_2 gain of 6

CH_4 gain of 8

Thus the ratio of oxidation number changes for C : CH_2 : CH_4 is 4 : 6 : 8, or 2 : 3 : 4.

If coal were taken to be CH, the ratio would be 5 : 6 : 8.

Green Chemistry Questions

1. (a) 2. Alternative reaction conditions for green chemistry.

   (b) 1. Prevention of waste.

   5. The use of auxiliary substances (e.g., solvents, separation agents, etc.) should be made unnecessary whenever possible, and should be innocuous when they are used.

   7. A raw material feedstock should be renewable rather than a depleting one whenever technically and economically practical.

   9. Catalytic reagents (as selective as possible) are superior to stoichiometric reagents.

   10. Chemical products should be designed so that at the end of their function they do not persist in the environment but break down into innocuous degradation products.

2. Made from annually renewable resources (corn, sugar beets; eventually waste biomass will be used).

   Production of PLA consumes 20–50% less fossil fuel resources than petroleum-based polymers.

Uses natural fermentation to produce lactic acid, uses no organic solvents or other hazardous substances in the production of PLA.

Uses catalysts, resulting in reduced energy consumption and resource consumption.

High yields of >95% are obtained.

Use of recycle streams help to reduce waste.

PLA can be recycled (converted back to monomer via hydrolysis, then repolymerized to produce virgin polymer), i.e., closed-loop recycling.

PLA can be composted (biodegradable); complete degradation occurs in a few weeks under normal composting conditions.

3. Growing crops, whether they are used to produce food or chemicals, require fertilizers and pesticides. Energy is needed to plant, cultivate and harvest; to produce, transport, and apply fertilizers and pesticides; to make and run tractors; to transport seeds, biomass, monomers, and polymers. Use of land to produce crops for chemicals also removes land that could be used to produce food and animal feed.

Additional Problems

1. Assuming the growth was exponential, then we have

$$E / E_0 = e^{kt}$$

Thus $1.40 = e^{20k}$ since the time period is (2000 – 1980) = 20 years.

Taking the ln of both sides of the equation, we have

$$\ln (1.40) = 20k$$

or $0.336 = 20k$, so $k = 0.017$.

Thus the average growth rate over the period was 1.7%.

From 1980 to 1990, the rate would have grown by

$$E / E_0 = e^{kt} = e^{0.017\times10} = e^{0.17} = 1.18$$

The energy use would have grown by 18% by 1990 (relative to 1980).

2. A doubling of energy E means that, at time t, $E = 2\,E_0$. Substituting for E in the equation

$E = E_0 e^{kt}$

Then $2\,E_0 = E_0 e^{kt}$

So $e^{kt} = 2$

Taking the natural logarithm of both sides, then

$kt = \ln(2) = 0.69$

So $t = 0.69 / k$

Thus the doubling time is $0.69 / k$. By substitution of the k values of 0.04, 0.03, 0.015, and 0.010 (corresponding to 4%, 3%, 1.5%, and 1% annual growth rates), we obtain doubling times of 17, 23, 46, and 69 years, respectively.

3. By definition, olefins contain free C=C bonds, which makes them more reactive in creating smog than aromatics because such bonds are more highly susceptible to attack by the OH free radical. Alkanes are quite unreactive because they contain no C=C or C=O bonds.

4. Consider the release of an equal amount of heat by the burning of oil—i.e., polymeric CH_2—and of natural gas; thus, the amounts of O_2 consumed are equal:

$$4\,CH_2 + 6\,O_2 \longrightarrow 4\,CO_2 + 4\,H_2O$$
$$3\,CH_4 + 6\,O_2 \longrightarrow 3\,CO_2 + 6\,H_2O$$

Thus, combustion of three moles of CH_4 "saves" 1 mole of CO_2, while still producing the same heat as burning oil. But the release of every one mole of CH_4 is equivalent to releasing 23 moles of CO_2, so releasing 1/23 mole of CH_4 would produce the equivalent greenhouse effect as one mole of CO_2.

Therefore, you cannot release more than 1/23 mole CH_4 when burning 3 moles of it, so the fraction (maximum) of release is:

$$\frac{1/23}{3 + 1/23} = 0.014 \quad \text{i.e., 1.4\% is the maximum release}$$

5. 5 metric tonnes = 5000 kg = 5.0×10^6 g

If this was converted to dry ice, the volume would be $5.0 \times 10^6 \text{ g} / 1.56 \text{ g cm}^{-3} = 3.2 \times 10^6 \text{ cm}^3$

The equation relating volume to radius of a sphere is: $V = 4/3\, \pi r^3$

$\therefore r = (3V / 4\pi)^{1/3} = (3 \times 3.2 \times 10^6 \text{ cm}^3 / 4 \times 3.142)^{1/3} = (7.65 \times 10^5 \text{ cm}^3)^{1/3} = 91.5 \text{ cm}$

$\therefore d = 183 \text{ cm} = 1.83 \text{ m}$

CHAPTER 7

Biofuels and Other Alternative Fuels

Problem 7-1

The average oxidation number of the carbon atoms in glucose, $C_6H_{12}O_6$, is determined by the fact that 12 times the +1 O.N. of H plus six times the O.N. of oxygen of –2 add up to zero, the net charge on the molecule. Hence the O.N. for the carbons in glucose must be zero.

In ethanol, the oxidation numbers of the hydrogen and oxygen atoms add up to + 6 – 2 = + 4, so the carbon atoms must average –2 each here.

Thus the carbon atoms in glucose that are converted to ethanol do become more reduced. However the O.N. of carbon in each CO_2 molecule is +4 (since each O is –2), so overall the carbon atoms suffer no net change in O.N. in being converted to ethanol and CO_2 since four carbons are reduced from 0 to –2 and two carbons are oxidized from 0 to +4.

Problem 7-2

Since the molar masses of methanol, CH_3OH, is 32.05 g, and that for ethanol, C_2H_5OH, is 46.08 g, the amounts of heat released by combustion per gram are 726 / 32.05 = 22.7 kJ and 1367 / 46.08 = 29.7 kJ, respectively. Multiplying both these values of kJ / g by the density of 0.79 g / mL gives energy releases of 17.9 and 23.5 kJ / mL, respectively.

Thus, ethanol is superior to methanol in energy intensity on both a mass and a volume basis. Since, according to the text, methane releases 55.6 kJ / gram and gasoline 43 kJ / gram, both are superior to the alcohols on that basis.

Problem 7-3

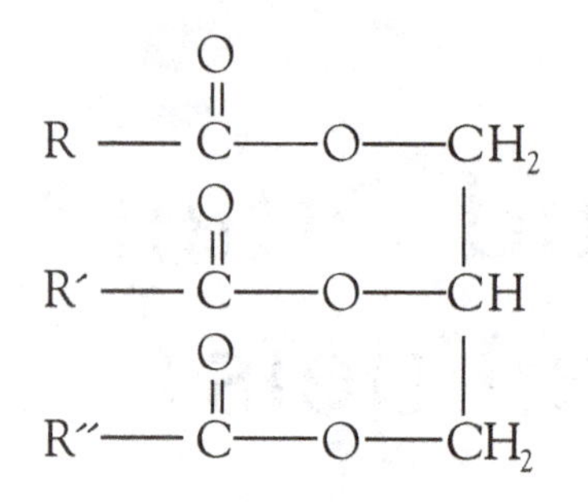

Problem 7-4

Since the process is exothermic, some of the fuel value of the biofuel is released during the hydrogenation process. Thus the heat released when the modified biofuel is combusted will be less, a disadvantage.

Problem 7-5

$$CH_3—CH_2—CH_2—CH_2OH$$

$$CH_3—CH_2—\underset{\displaystyle OH}{\underset{|}{CH}}—CH_3$$

$$(CH_3)_2CH—CH_2OH$$

$$(CH_3)_2\underset{\displaystyle OH}{\underset{|}{C}}—CH_3$$

Problem 7-6

The reaction is:

$$2\,H_2 + CO \longrightarrow CH_3OH$$

$$\begin{aligned} \text{so } \Delta H &= \Delta H_f\,(CH_3OH) - 2\,\Delta H_f\,(H_2) - \Delta H_f\,(CO) \\ &= -239.1 - 2 \times 0 - (-110.5) \\ &= -128.6 \text{ kJ mol}^{-1} \end{aligned}$$

The reaction is exothermic, so heat is a product along with methanol. Thus by Le Chatelier's Principle, decreasing the temperature will drive the equilibrium toward the right, increasing the product yield. Thus to obtain maximum equilibrium yield, it is important to carry out the process at as low a temperature as possible. Since reactions tend to be slow at low temperatures, a low-temperature catalyst would be useful to achieve the favorable equilibrium quickly.

Problem 7-7

In the product CH_3O, the ratio of H to CO is 3:1, so the ratio of H_2 to CO in the reactants would be 1.5:1, or 3:2, because each H_2 supplies two hydrogens.

The reaction of methane with steam is:

$$CH_4 + H_2O \longrightarrow CO + 3H_2$$

so the H_2 to CO ratio is 3:1, i.e., too much H_2 is produced for CH_3O. To convert some H_2 to CO, we use the shift reaction written in the direction:

$$H_2 + CO_2 \rightleftharpoons H_2O + CO$$

	H_2	CO
Initial amt.	$3a$	a
At new eqm.	$3a - x$	$a + x$

Since we want the H_2/CO ratio at the new equilibrium to be 3:2, we have:

$$\frac{3a - x}{a + x} = \frac{3}{2}$$

By algebraic manipulation of this equation, we obtain:

$$6a - 2x = 3a + 3x$$

$$\text{or } -5x = -3a$$

so

$$\frac{x}{a} = \frac{3}{5} = 0.6$$

Since the initial amount of H_2 was $3a$, then dividing both sides of this equation by 3, we obtain the ratio of x, the amount of H_2 to be transformed, to $3a$, its original amount:

$$\frac{x}{3a} = 0.20$$

Thus 20% of the original H_2 needs to be converted to CO.

Problem 7-8

Since the molar masses of methanol, CH_3OH, is 32.05 g, and that for ethanol, C_2H_5OH, is 46.08 g, the amounts of heat released by combustion per gram are 726 / 32.05 = 22.7 kJ and 1367 / 46.08 = 29.7 kJ, respectively. Multiplying both these values of kJ / g by the density of 0.79 g / mL gives energy releases of 17.9 and 23.5 kJ / mL, respectively.

Thus, ethanol is superior to methanol in energy intensity on both a mass and a volume basis. Since, according to the text, methane releases 55.6 kJ / gram and gasoline 43 kJ / gram, both are superior to the alcohols on that basis.

Problem 7-9

Since the energy E required is 285.8 kJ / mole, then using the formula in Chapter 1 we can solve for the wavelength λ:

$$E = 119,627 / \lambda$$
$$\text{so } \lambda = 119,627 / E = 119,627 \text{ kJ mole}^{-1} \text{ nm} / 285.8 \text{ kJ mole}^{-1}$$
$$= 419 \text{ nm}$$

This wavelength lies in the visible region, although water does not absorb light at this wavelength to a significant extent.

Problem 7-10

First, obtain the balanced equation for each combustion reaction:

$$H_2 + 1/2\, O_2 \longrightarrow H_2O$$
$$CH_4 + 2\, O_2 \longrightarrow CO_2 + 2\, H_2O$$

Since one mole of CH_4 consumes four times as much oxygen as does one mole of H_2, the ratio of the heat released is also about 4:1.

Problem 7-11

Since the molar mass of H_2 is 2.02 g, its heat of combustion is 242 kJ mole^{-1} / 2.02 g mole^{-1} = 120 kJ per gram, which is clearly superior on a weight basis to that of methane of 55.6 kJ per gram (all data taken from the text). Per mole of gas, and hence per molar volume, methane is clearly superior, at 890 kJ / mole compared to 242 kJ / mole for hydrogen.

Problem 7-12

For an ideal gas, we have PV = nRT.

Thus the molar density $n / V = P / RT = 700 \text{ atm} / (0.082 \text{ L atm mole}^{-1} \text{ K}^{-1} \times 300 \text{ K})$
$= 28.46 \text{ moles} / \text{L}$

Converting to mass and a volume of 1 m^3,

$$(28.46 \text{ moles} / 1 \text{ L}) \times (2.02 \text{ g} / 1 \text{ mole}) \times (1000 \text{ L} / 1 \text{ m}^3) = 57,480 \text{ g m}^{-3}$$

Thus, the ideal gas density under these conditions would be 57.48 kg m^{-3}. Since the actual density is only 37 kg m^{-3}, the error in the calculated density is

$$(57.48 - 37) / 37 \times 100\% = 55\%$$

Problem 7-13

The mole ratio of Ti to H_2 in TiH_2 is 1:1; thus, for one mole (2.02 g) of H_2, the required mass of titanium is its atomic mass, 47.88 g. Since each kilogram of H_2 contains 1000 g / 2.02 g mole H_2 = 495 moles H_2, the required mass of titanium for one kilogram of hydrogen is

$$495 \times 47.88 \text{ g} = 23.7 \text{ kg}$$

By the same reasoning, the mass of magnesium is 12.0 kg; thus, from a weight standpoint, magnesium is superior to titanium with respect to its ability to store hydrogen.

Problem 7-14

Ammonia borane has the formula BH_3NH_3, so its

Molar mass = 1 × 10.81 + 6 × 1.01 + 1 × 14.01 = 30.88

The percentage of hydrogen would then be 6 × 1.01 / 30.88 × (100%) = 19.6%.

The empirical formula for the BNH polymer gives a molar mass of 10.81 + 14.01 + 1.01 = 25.83, in which the percent of hydrogen is 1.01 / 25.83 × (100%) = 3.9%.

Thus the loss of hydrogen in forming the polymer is 19.6 – 3.9 = 15.7%. Both the hydrogen content of ammonia borane and the release of it upon forming the polymer fit within the DOE guideline of at least 6% hydrogen by weight, but the polymer itself does not.

Note that the equation for polymer formation from ammonia borane does not balance; it is not clear whether the excess hydrogen not accounted for in the equation would be available as a gas or not.

Problem 7-15

The first half-reaction mentioned is (unbalanced).

$$O_2 + H_2O + e^- \longrightarrow OH^-$$

This can be balanced by placing a 2 in front of H_2O, and then a 4 in front of OH^- and therefore a 4 in front of e^- to balance charge:

$$O_2 + 2\,H_2O + 4\,e^- \longrightarrow 4\,OH^-$$

The second half-reaction is described as:

$$OH^- + H_2 \longrightarrow H_2O + e^-$$

The balanced reaction can be obtained by doubling the amounts of OH^- and H_2O, and having two electrons produced:

$$2\,OH^- + H_2 \longrightarrow 2\,H_2O + 2\,e^-$$

The balanced overall reaction must have as many electrons on the right as on the left; thus we double this half-reaction before adding it to that above. The addition gives:

$$O_2 + 2\,H_2O + 4\,e^- + 4\,OH^- + 2\,H_2 \longrightarrow 4\,OH^- + 4\,H_2O + 4e^-$$

Canceling common terms, we obtain:

$$O_2 + 2\,H_2 \longrightarrow 2\,H_2O$$

Problem 7-16

From the description in the text, the hydrogen half-reaction is:

$$H_2 + CO_3^{2-} \longrightarrow CO_2 + H_2O + e^-$$

This half-reaction is already balanced except for charge, so we obtain:

$$H_2 + CO_3^{2-} \longrightarrow CO_2 + H_2O + 2\,e^-$$

From the description in the text, the other half-reaction is:

$$CO_2 + e^- + O_2 \longrightarrow CO_3^{2-}$$

We need a 2:1 ratio of CO_2 to O_2 to balance the oxygen, and two electrons on the left to balance charge for each carbonate on the right, so we obtain:

$$2\,CO_2 + 4\,e^- + O_2 \longrightarrow 2\,CO_3^{2-}$$

If these equations are to be added, the first must be multiplied by 2 if the electrons are to be equal on the two sides; thus we obtain:

$$2\,H_2 + 2\,CO_3^{2-} + 2\,CO_2 + 4\,e^- + O_2 \longrightarrow 2\,CO_2 + 2\,H_2O + 4\,e^- + 2\,CO_3^{2-}$$

After canceling common terms, we have:

$$2\,H_2 + O_2 \longrightarrow 2\,H_2O$$

Green Chemistry Problems

1. (a) 3. The design of chemicals that are less toxic than current alternatives or inherently safer with regard to accident potential.

   (b) 1. The prevention of waste.

   3. Wherever practicable, synthetic methodologies should be designed to use and generate substances that possess little or no toxicity to human health and the environment.

   6. Energy reduction.

   7. A raw material feedstock should be renewable rather than depleting whenever technically and economically practical.

2. The use of renewable feedstocks, reduction in energy, and reduction of pollutants.

3. (a) 1. The use of alternative synthetic pathways.

   (b) 1. The prevention of waste.

   7. A raw material feedstock should be renewable rather than depleting whenever technically and economically practical.

4. The reduction of the use of fossil fuels and the reduction of greenhouse gas emissions.

Additional Problems

1. Wood is a renewable resource, so the CO_2 emitted upon burning it will be reabsorbed in the growing of trees to replace those harvested. Hence, in principle at least, burning wood yields no net CO_2 emissions, whereas those emitted by burning of fossil fuels are not reabsorbed because new sources of them are being created.

2. (a) Methanol's combustion would produce CO_2 and water:

$$CH_3OH + O_2 \longrightarrow CO_2 + H_2O$$

After balancing, we obtain:

$$CH_3OH + 1.5\,O_2 \longrightarrow CO_2 + 2\,H_2O$$

or

$$2\,CH_3OH + 3\,O_2 \longrightarrow 2\,CO_2 + 4\,H_2O$$

The ratio of 1.5 of molecular oxygen to CO_2 is the same as that for CH_2— i.e., oil—so it is more similar to oil than to natural gas or coal in terms of energy released (which is proportional to O_2 consumed) per CO_2.

(b) The unbalanced reaction is:

$$C + H_2O \longrightarrow CH_3OH + CO_2$$

One way to balance this reaction is to realize that in the reactant, carbon has a zero oxidation number, whereas in the products methanol and carbon dioxide it has the –2 and +4 numbers respectively. Thus twice as much methanol as carbon dioxide is produced to balance the oxidation numbers; the balanced equation follows readily:

$$3\,C + 4\,H_2O \longrightarrow 2\,CH_3OH + CO_2$$

(c) If we combine the equations producing and combusting 2 CH_3OH, then after cancellation of common terms we obtain:

$$3\,C + 3\,O_2 \longrightarrow 3\,CO_2$$

$$\text{i.e., } C + O_2 \longrightarrow CO_2$$

Thus just as much CO_2 is produced this way as is obtained by simply burning it; the conversion to methanol is just another way of performing the same overall reaction, and so its energy output must be exactly the same (Hess' Law).

3. The reaction of coal (taken as carbon, C) with steam produces a 1:1 ratio of H_2 to CO; as in the example in the text, we shall assume that the amounts produced of each are the quantity a. The shift reaction, written so as to produce more hydrogen, is then:

	CO	+	H_2O	$\rightleftharpoons$	CO_2	+	H_2
From coal	a						a
After rxn.	a – x						a + x

We require the ratio (a + x) / (a – x) to equal 2, so it follows that

$$a + x = 2a - 2x$$

or, rearranging to solve for x, we obtain

$$x = a / 3.$$

Thus one-third of the CO produced from coal must be converted to additional H_2.

The reactions to add together then are:

$$C + H_2O \rightleftharpoons CO + H_2$$

$$1/3\,[CO + H_2O \rightleftharpoons CO_2 + H_2]$$

$$2/3\,[CO + 2\,H_2 \rightleftharpoons CH_3OH]$$

Overall we obtain

$$C + 4/3\,H_2O \rightleftharpoons 2/3\,CH_3OH + 1/3\,CO_2$$

$$\text{or } 3\,C + 4\,H_2O \rightleftharpoons 2\,CH_3OH + CO_2$$

4. Synthesis gas is formed here by a combination of methane and carbon dioxide. We use equal volumes, and hence equal numbers, of moles of each reactant, and so the balanced equation must be:

$$CH_4 + CO_2 \longrightarrow 2\,CO + 2\,H_2$$

For this reaction,

$$\begin{aligned}\Delta H &= 2\,\Delta H_f(CO) + 2\,\Delta H_f(H_2) - \Delta H_f(CH_4) - \Delta H_f(CO_2)\\ &= 2 \times (-110.5) + 2 \times 0 - (-74.9) - (-393.5)\\ &= +247.4\ \text{kJ mol}^{-1}\end{aligned}$$

Since the reaction is endothermic, increasing the temperature will drive it to the right, i.e., toward greater production of CO and H_2.

Since the number of moles of gas is greater in the product than in the reactants (4 vs. 2), increasing the pressure will drive the reaction back to the reactants. Thus *low* pressure will be favorable to formation of products.

To produce methanol from synthesis gas, we require a 2:1 ratio of H_2 to CO. Since the reaction between CH_4 and CO_2 produces a 1:1 ratio, as does that between coal and steam, we must convert one-third of the reactant CO to H_2. Thus we have:

$$\begin{aligned}&CH_4 + CO_2 \longrightarrow 2\,CO + 2\,H_2\\ &1/3\,[2\,CO + 2\,H_2O \longrightarrow 2\,CO_2 + 2\,H_2]\end{aligned}$$

$$\text{net}\quad CH_4 + 1/3\,CO_2 + 2/3\,H_2O \longrightarrow 4/3\,CO + 8/3\,H_2$$

This amount of synthesis gas will produce 4/3 moles of methanol

$$4/3\,CO + 8/3\,H_2 \longrightarrow 4/3\,CH_3OH$$

to give a net reaction in the production of methanol as:

$$CH_4 + 1/3\,CO_2 + 2/3\,H_2O \longrightarrow 4/3\,CH_3OH$$

Combustion of this methanol produces 4/3 mole of CO_2:

$$4/3\,CH_3OH + 2\,O_2 \longrightarrow 4/3\,CO_2 + 8/3\,H_2O$$

Thus the net process of production and combustion of methanol produced in this way is:

$$CH_4 + 2\,O_2 \longrightarrow CO_2 + 2\,H_2O$$

In other words, the reaction is identical to that of simply burning methane, and there is no net decrease in the amount of CO_2 that is added to the air. Thus methanol produced in this manner is *not* "renewable."

5. The unbalanced reaction for the production of hydrogen and carbon dioxide from methane gas using steam is presumably

$$CH_4 + H_2O \longrightarrow H_2 + CO_2$$

Balancing the oxygen atoms, then the hydrogen, gives the balanced equation

$$CH_4 + 2\,H_2O \longrightarrow 4\,H_2 + CO_2$$

Hydrogenation of CH to give CH_2 corresponds to

$$CH + \tfrac{1}{2}H_2 \longrightarrow CH_2$$

If we multiply this reaction by 8 so as to use all the hydrogen produced in the previous reaction, and then add the reactions together, we obtain the overall reaction:

$$8\,CH + CH_4 + 2\,H_2O \longrightarrow 8\,CH_2 + CO_2$$

CHAPTER 8

Renewable Energy Technologies: Hydroelectric, Wind, Solar, Geothermal, and Marine Energy and Their Storage

Problem 8-1

a) Since wind energy rate E is proportional to wind speed v, we can compare the relative amounts of energy at two different speeds from the ratio:

$$E_2 / E_1 = v_2^3 / v_1^3 = (v_2 / v_1)^3$$

By substitution of the wind velocity values, we obtain

$$E_2 / E_1 = (8 / 5)^3 = 4.1$$

Thus the energy at the higher speed is 4.1 times that at the lower speed.

b) Using the same logic as in part (a), doubling the speed increases the energy yield by a factor of $2^3 = 8$.

c) Using the relationship $E = kv^3$, where k is a constant, then the energy rate E at 12 mps is $k(12)^3 = 1728k$, whereas that at 5 mps, the energy is $k(5)^3 = 125k$. Since the higher velocity occurs only 10% of the time, the total energy it delivers is $0.10 \times 1728k = 172.8k$. The energy from the lower speed 90% of the time is $0.9 \times 125k = 112.5k$. Thus the faster speed delivers more energy, even though it is available only 10% of the time.

Problem 8-2

The unbalanced decomposition reaction would be:

$$CeO_2 \longrightarrow Ce_2O_3 + O_2$$

which is easily balanced to give:

$$2\,CeO_2 \longrightarrow Ce_2O_3 + \tfrac{1}{2}\,O_2$$

The reaction of steam with Ce_2O_3 would reform CeO_2 and liberate hydrogen gas:

$$Ce_2O_3 + H_2O \longrightarrow H_2 + 2\,CeO_2$$

Adding the last two reactions together and cancelling common terms, we obtain:

$$H_2O \longrightarrow H_2 + \tfrac{1}{2}\,O_2$$

Problem 8-3

Since the fraction that can be converted is given by the ratio $(T_h - T_c) / T_h$ and because $T_h = 900 + 273 = 1173$ K, and $T_c = 100 + 273 = 373$ K, thus the ratio is:

$$(1173 - 373) / 1173 = 0.68$$

Thus 68% of the heat can be converted.

Problem 8-4

The maximum fractional conversion is $(T_h - T_c) / T_h$. Here the difference $T_h - T_c$ is 20 degrees (Celsius or Kelvin), and T_h is $273 + (25 + 20) = 318$ K, so the ratio is:

$$20/318 = 0.063$$

Thus 6.3% of the energy associated with the 20 degree gradient can be converted to electricity.

Problem 8-5

Here the fractional conversion is 0.50, and the value of T_c is $273 + 57 = 330$ K, so because the fraction is $(T_h - T_c) / T_h$, thus we have:

$$0.50 = (T_h - 330) / T_h$$

Multiplying both sides by T_h, we obtain:

$$0.5\,T_h = T_h - 330$$
$$-0.5\,T_h = -330$$
$$\text{so } T_h = 660$$

Thus the Celsius temperature to which the heat source must be raised is $660 - 273 = 387$°C.

Additional Problems

1. First calculate the surface area of the Earth, in square meters:

$$\begin{aligned} S &= 4\pi r^2 \\ &= 4 \times 3.14 \times (6400 \times 10^3\,\text{m})^2 \\ &= 5.1 \times 10^{14}\,\text{m}^2 \end{aligned}$$

Thus the total watts received on average is:

$$\frac{342\ \text{watts}}{1\ \text{m}^2} \times 5.1 \times 10^{14}\,\text{m}^2 = 1.76 \times 10^{17}\ \text{watts}$$

Since 1 watt = 1 joule second, the Earth receives 1.7×10^{17} joules per second. To calculate this energy on a yearly basis, we multiply by the number of seconds in a year:

$$1\ \text{year} \times \frac{365\ \text{d}}{1\ \text{y}} \times \frac{24\ \text{h}}{1\ \text{d}} \times \frac{60\ \text{min}}{1\ \text{h}} \times \frac{60\ \text{sec}}{1\ \text{min}} = 3.15 \times 10^7\ \text{sec}$$

Since $1.76 \times 10^{17} \times 3.15 \times 10^7 = 5.6 \times 10^{24}$, the total energy received per year is 5.6×10^{24} J.

The ratio of the energy use of 500 EJ, that is, 500×10^{18} J, to the total received is $500 \times 10^{18} / 5.6 \times 10^{24} \times 100\% = 0.009\%$.

2. The constant $c^2 = (4/\pi)\, v_{avg}^{\ 2} = 1.273\, v_{avg}^{\ 2}$

So for $v_{avg} = 5$, $c^2 = 31.83$

Hence, for $v_{avg} = 5$, the function $F = 0.03141\, v^4 \exp(-0.03141\, v^2)$,

whereas for $v_{avg} = 10$, the function $F = 0.007854\, v^4 \exp(-0.007854\, v^2)$.

The plot at the lower average wind speed reaches a maximum at about 8 meters per second, where the function has the value 17.2, and that for the higher speed maximizes at about 16 m s^{-1}, where F is 68.9.

Clearly by comparing the plots (see below) of the energy F versus v, the total energy that can be collected (area under the curve) at lower average wind speed is much less than that at the higher speed.

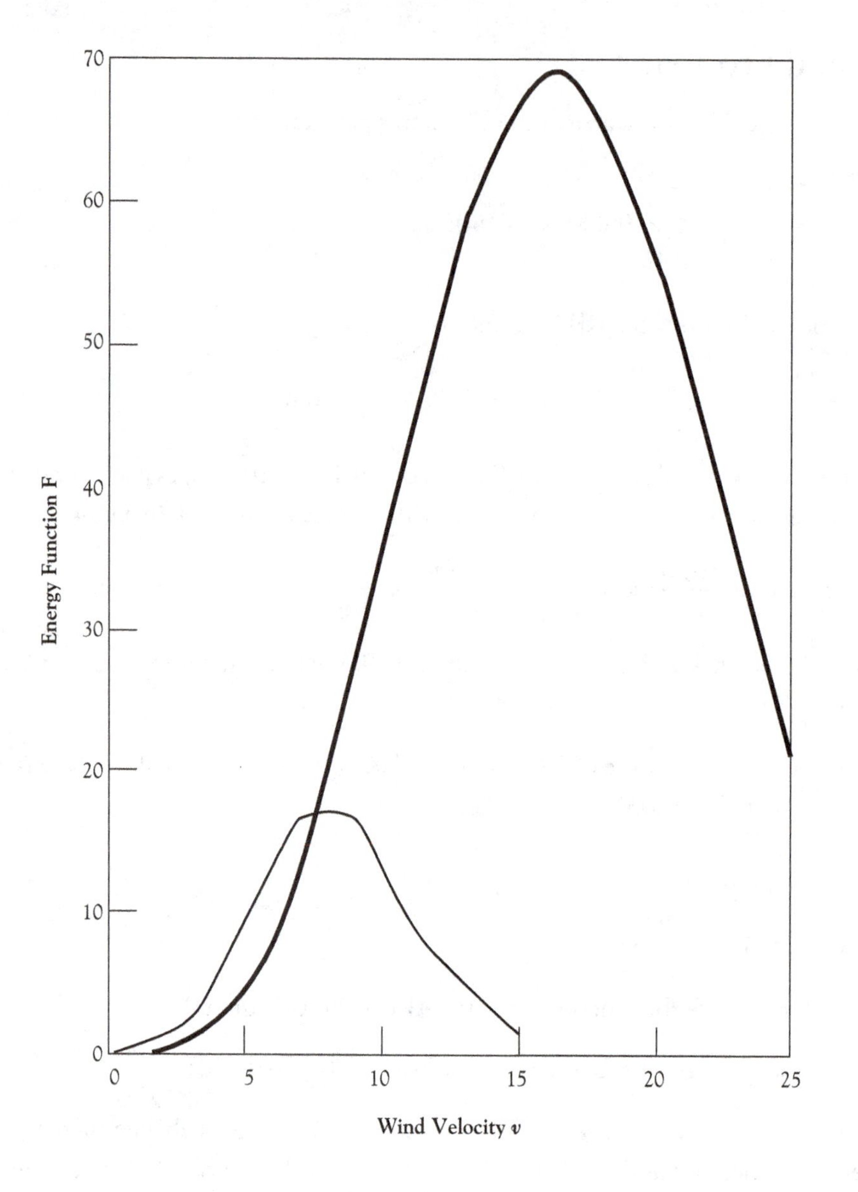

3. The values and logarithms are given in the table below:

Year	Power capacity	ln of power capacity
1996	6.1	1.808
1998	10.0	2.303
2000	17.4	2.856
2002	31.3	3.444
2004	47.6	3.863
2006	74.6	4.312
2008	121	4.796
2010	198	5.288

From the best straight line through the plot of ln (power) versus year shown below, the capacity predicted for 2015 is about 665 (because the projected ln is 6.50), and that for 2020 is about 2300 (because the projected ln is about 7.75).

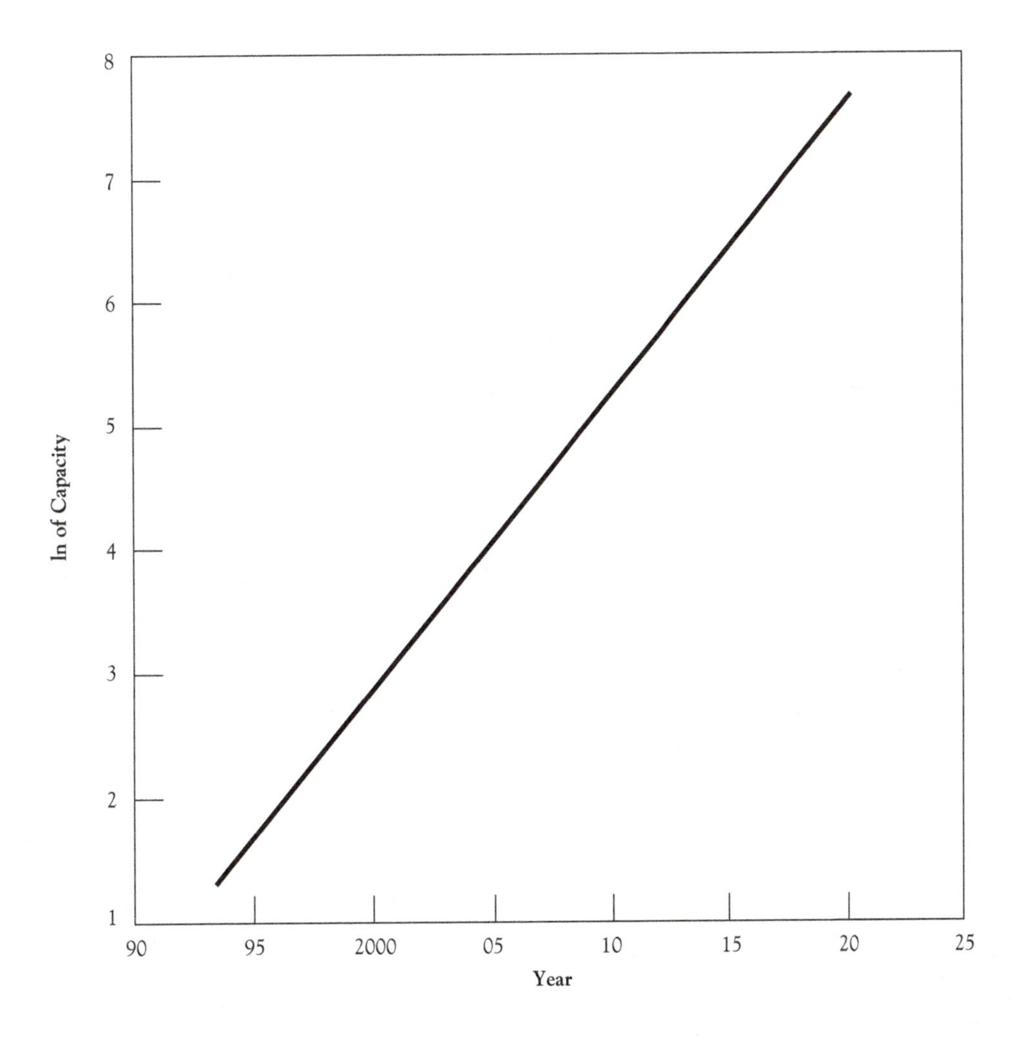

CHAPTER 9

Radioactivity, Radon, and Nuclear Energy

Problem 9-1

Recall that both the total mass number (sum of leading superscripts) and the total number of protons (sum of leading subscripts) are identical on the two sides of the equation.

(a) For "blank" species, mass number must be 222 – 4 = 218
For "blank" species, number of protons must be 86 – 2 = 84, so

"blank" species is ${}^{218}_{84}Po$ since element 84 is Po (see Periodic Table).

(b) Since β is equivalent to ${}^{0}_{-1}e$, then for "blank" species

mass number must be 214 – 0 = 214, and
the number of protons must be 83 – (–1) = 84 so

"blank" species is ${}^{214}_{84}Po$ since element 84 is Po (see Periodic Table).

(c) Since Po is element 84 and Pb is element 82 (see Periodic Table), the reaction can be rewritten as

$${}^{214}_{84}Po \longrightarrow {}^{210}_{82}Pb + \_\_\_\_\_\_\_\_$$

Thus, for the "blank" species,

mass number must be 214 – 210 = 4
the number of protons must be 84 – 82 = 2 (i.e., helium)

"blank" species is ${}^{4}_{2}He$ = α particle

(d) Since an α particle is ${}^{4}_{2}He$, the reaction can be rewritten as

$$\_\_\_\_\_\_\_\_ \longrightarrow {}^{234}_{90}Th + {}^{4}_{2}He$$

So, for the "blank" species,

mass number must be 234 + 4 = 238
the number of protons must be 90 + 2 = 92 (i.e., uranium)

Thus, "blank" species is $^{238}_{92}U$

Problem 9-2

From the value of the half-life quoted in the text, i.e., 4.5×10^9 y, we can calculate the value of k:

$k = 0.693 / t_{1/2} = 0.693 / 4.5 \times 10^9 \text{ y} = 1.54 \times 10^{-10} \text{ y}^{-1}$

When 99% of the uranium has disintegrated, the fraction f that still remains is 1.00 – 0.99 = 0.01.

For first-order decay, we know that $f = e^{-kt}$. Taking the ln of both sides, then

$\ln (f) = -kt$

$$\text{so } t = -\ln (f) / k = -\ln (0.01) / 1.54 \times 10^{-10} \text{ y}^{-1}$$

$$= -(-4.605) / 1.54 \times 10^{-10} \text{ y}^{-1}$$

$$= 3.0 \times 10^{10} \text{ years}$$

Thus, it takes 30 billion years for 99% of the uranium to decay.

Problem 9-3

According to the equation quoted in Chapter 8, the maximum fraction f of heat that can be converted to electricity is $(T_{hot} - T_{cold}) / T_{hot}$.

In the current example, $T_{hot} = 300 + 273 = 573$ K, and $T_{cold} = 17 + 273 = 290$ K.

Thus $f = (573 - 290) / 573 = 0.49$.

Repeating the calculation for $T_{hot} = 550 + 273 = 823$ K, we obtain f = 0.65.

The calculated values of 49% and 65% lie substantially above the observed efficiencies of 30% and 40% for nuclear and coal-fired power plants.

Problem 9-4

We recognize that $128 = 2^7$, so $1/128 = (1/2)^7$.

Thus 7 half-life periods are required for plutonium to decrease to 1/128 of its original value.

Time required = 7 × 24,000 y = 168,000 years

Problem 9-5

Since deuterium is $^{2}_{1}H$, the first reaction is

$$^{2}_{1}H \quad + \quad ^{3}_{2}He \longrightarrow \, ^{4}_{2}He \quad + \quad ^{1}_{1}p \text{ (or } ^{1}_{1}H)$$

and the second is

$$2\, ^{3}_{2}He \longrightarrow \, ^{4}_{2}He \quad + \quad 2\, ^{1}_{1}p \text{ (or } 2\, ^{1}_{1}H)$$

Problem 9-6

The symbol for tritium is $^{3}_{1}H$ and that for a neutron is $^{1}_{0}n$, so the reaction is

$$^{6}_{3}Li \quad + \quad ^{1}_{0}n \longrightarrow \, ^{3}_{1}H \quad + \quad ^{4}_{2}He$$

Note that the mass numbers on the two sides balance (6 + 1 = 3 + 4), as do the atomic numbers (3 + 0 = 1 + 2).

Problem 9-7

To calculate the energy release, we must first determine the change in mass in going from reactants to products. After cancelling 1 neutron from both sides, the net reactant is just one ^{235}U nucleus, mass = 235.044 g / mol, and the mass of the products is the sum of that for ^{142}Ba + ^{91}Kr + 2 neutrons = 141.916 + 90.923 + 2 × 1.008665 = 234.856 g / mol. Thus the change in mass is 235.044 – 234.856 = 0.188 grams.

We obtain the energy E from the mass change, m, and the speed of light, c:

$$E = mc^2 = 0.188 \times 10^{-3} \text{ kg} \times (2.99792 \times 10^8 \text{ m / s})^2$$

$$= 1.69 \times 10^{13} \text{ kg m}^2 / \text{s}^2 = 1.69 \times 10^{13} \text{ J}$$

Dividing this value of E by the mass of the reactants (a U nucleus and a neutron) of 236.05 gives 7.16×10^{10} J / gram, which is only about one-fifth that of 1.69×10^{12} / (2.0140 + 3.01605) = 3.36×10^{11} J / gram from the fusion of deuterium and tritium.

Box 9-1, Problem 1

Applying the same analysis to the variation in [B] with time as in the Box for [C] gives the expression for the ratio of steady-state concentrations in terms of half-lives:

$$[B]_{ss} / [A]_{ss} = t_B / t_A$$

Thus we obtain an expression for the steady-state concentration of B:

$$[B]_{ss} = [A]_{ss}\, t_B / t_A$$

Substituting this expression into the ratio for $[C]_{ss} / [B]_{ss}$ in the Box gives

$$[C]_{ss} / [A]_{ss} = t_C / t_A$$

This general result can be obtained for any member of the series, thus proving that the steady-state concentration of any member is proportional to its disintegration half-life.

Additional Problems

1. Since the number of protons *decreases* upon positron emission, the atomic number also **decreases** by one, i.e., the element is converted to the one just before it in the periodic table.

   The symbol for a positron is ${}^{0}_{1}e$, since the atomic number of the original nucleus decreases by 1, but the mass number remains unchanged.

   Balanced equations: $${}^{22}_{11}\text{Na} \longrightarrow {}^{22}_{10}\text{Ne} + {}^{0}_{1}e$$

   $${}^{13}_{7}\text{N} \longrightarrow {}^{13}_{6}\text{C} + {}^{0}_{1}e$$

2. The text gives the half-life of ^{131}I as 8 days.

   $$t_{1/2} = 0.693 / k \qquad \therefore k = 0.693 / t_{1/2} = 0.693 / 8 \text{ day} = 0.0866 \text{ day}^{-1}$$

   After 7 days, $f = e^{-kt} = e^{-0.0866 \text{ day}^{-1} \times 7 \text{ days}} = 0.545$

   This is the fraction remaining; thus the percentage remaining is 54.5 %.

   The time required for 99% of the ^{131}I to decay is equal to the time required for 1% to remain:

   $$f = 0.01 = e^{-kt} = e^{-0.0866 / \text{day} \times t}$$

   Taking the natural log of both sides, and solving for t, we obtain

   $$\therefore t = -\ln(0.01) / 0.0866 \text{ day}^{-1} = 53.2 \text{ days.}$$

3. We need the molar masses; we assume the masses of pure isotopes equal their mass numbers:

   $^{235}UF_6$: $235 + 6 \times 19.0 = 349$ atomic mass units

   $^{238}UF_6$: $238 + 6 \times 19.0 = 352$ atomic mass units

   $\therefore$ rate 235 / rate 238 $= \sqrt{352} / \sqrt{349} = 1.0043$

   Thus, the effusion rate of $^{235}UF_6$ is just 0.43% faster than that of $^{238}UF_6$. Thus, very little enrichment would occur in a single step, making it necessary to repeat the process hundreds of times to increase the fraction of ^{235}U from its natural abundance of 0.7% to the level of 3.0% required for its use in nuclear power plants.

CHAPTER 10

The Chemistry of Natural Waters

Problem 10-1

(a) Convert volume to mass of solution by using density = 1.0 g / mL, so

$$1\ \text{L} = 1000\ \text{mL} = 1000\ \text{g solution}$$

Thus the concentration is 0.04×10^{-6} g pollutant / 1000 g of solution. To obtain the value on the ppm scale, the mass of solution must be 10^6 g, so multiply top and bottom by 1000.

$$\begin{aligned}\text{Solution: } & 0.04 \times 10^{-3}\ \text{g pollutant} / 10^6\ \text{g solution} \\ & = 0.04 \times 10^{-3}\ \text{ppm} \\ & = 4.0 \times 10^{-5}\ \text{ppm}\end{aligned}$$

To obtain concentration on the ppb scale, multiply top and bottom of ppm concentration by a further 1000, giving a result of 0.04 ppb.

(b) $$3\ \text{ppb means } \frac{3\ \text{g pollutant}}{10^9\ \text{g solution}}$$

To convert denominator to a volume basis, recall 1000 g aqueous solution = 1000 mL = 1 L, because density is 1.0 g/mL.

$$\frac{3\ \text{g pollutant}}{10^9\ \text{g solution}} \times \frac{1000\ \text{g solution}}{1\ \text{L}} = 3.0 \times 10^{-6}\ \text{grams pollutant/L}$$

$$= 3.0\ \mu\text{g pollutant/L}$$

(c) The concentration in its original units is 0.30 μg solute / 1 g of solution, so it is already on the mass / mass basis used in *parts-per* scales in water. To determine the mass of solute in 1 billion grams of solution, multiply both numerator and denominator by 10^9:

$$0.30 \times 10^{-6}\ \text{g solute} \times 10^9 = 300\ \text{g solute}$$

Thus the concentration is 300 g solute / 10^9 g solution, i.e., 300 ppb.

Problem 10-2

From Henry's Law, $[O_2(aq)] = K_H P_{O_2}$

$$= 1.3 \times 10^{-3} \text{ M atm}^{-1} \times 0.21 \text{ atm}$$
$$= 0.00027 \text{ moles L}^{-1}$$

Converting to mass, because 1 mole O_2 = 32 g,

$$\therefore [O_2(aq)] = \frac{0.00027 \text{ moles } O_2}{1 \text{ L water}} \times \frac{32 \text{ g } O_2}{1 \text{ mole } O_2}$$

$$= 0.0087 \text{ g / L}$$
$$= 8.7 \text{ mg / L}$$

Problem 10-3

The 0°C solubility given in the text is 14.7 ppm. Since the equilibrium constant uses moles/L units, we must convert this solubility to such units.

$$\frac{14.7 \text{ g } O_2}{10^6 \text{ g } H_2O} \times \frac{1 \text{ mole } O_2}{32.0 \text{ g } O_2} \times \frac{1000 \text{ g } H_2O}{1 \text{ L } H_2O} = 4.59 \times 10^{-4} \text{ moles/L}$$

Using the defining equation

$$K_H = [O_2(aq)] / P_{O_2}$$

and given that $P_{O_2} = 0.21$ atm, then

$$K_H = 4.59 \times 10^{-4} / 0.21 = 2.2 \times 10^{-3} \text{ moles L}^{-1} \text{ atm}^{-1}$$

Problem 10-4

From Problem 10-2, we know that 1 L of water saturated with O_2 contains 0.00027 moles O_2, and this is capable of oxidizing 0.00027 mole of polymeric CH_2O via the balanced reaction

$$CH_2O + O_2 \longrightarrow CO_2 + H_2O$$

Since the molar mass of CH_2O is 30.0 g / mole, the mass of CH_2O is, therefore, 0.00027 mole $\times$ 30.0 g mole^{-1} = 0.0081 g, i.e., 8.1 mg.

Problem 10-5

The unbalanced redox reaction is

$$NH_3 + O_2 \xrightarrow{OH^-} NO_3^- + H_2O$$

Using standard redox balancing techniques, we obtain:

$$NH_3 + 2\,O_2 + OH^- \longrightarrow NO_3^- + 2\,H_2O$$

The water becomes less alkaline, since OH^- is consumed.

Problem 10-6

Since the oxidation number of oxygen is –2 and there are 7 of them, the total from all O's is –14. However, 2 of the 14 are accounted for by the charge on the ion, leaving 12 to be countered by the two Cr atoms. Thus, the oxidation number for each Cr is +6 here.

In the half-reaction, each Cr changes from +6 to +3, a gain of 3 electrons, so the total gain of electrons is 6, as in the balanced half-reaction. (The oxidation numbers of H and O do not change in the process.)

Problem 10-7

From the volume and molarity of the $Na_2Cr_2O_7$ solution, we can deduce the number of moles of it used in titration.

$$\text{Moles} = MV = \frac{0.0010 \text{ moles}}{\text{L}} \times 0.0083 \text{ L} = 8.3 \times 10^{-6} \text{ moles.}$$

Since (see text) the moles O_2 equal 6 / 4 times the moles of $Na_2Cr_2O_7$, then moles $O_2 = (6/4) \times 8.3 \times 10^{-6} = 1.25 \times 10^{-5}$ moles O_2. Since we require the mass of O_2, we multiply this value by its molar mass, 32.0 grams, to find that the amount of O_2 is 4.0×10^{-4} g, or 0.40 milligrams. Since the volume of the sample was 25 mL, or 0.025 L, the O_2 concentration is

$$0.40 \text{ mg} / 0.025 \text{ L} = 16 \text{ mg} / \text{L}$$

Problem 10-8

First convert the mass of O_2 into moles:

$$30 \times 10^{-3} \text{g } O_2 \times \frac{1 \text{ mole } O_2}{32.0 \text{ g } O_2} = 9.4 \times 10^{-4} \text{ moles } O_2$$

Thus the molar concentration of O_2 in the sample is 9.4×10^{-4} moles / L. The amount of O_2 in a 50 mL sample is therefore

$$0.050 \text{ L} \times 9.4 \times 10^{-4} \text{ moles } O_2 / \text{L} = 4.7 \times 10^{-5} \text{ moles } O_2$$

The required number of moles of $Na_2Cr_2O_7$ is (4 / 6) times this amount (see text), or 3.1×10^{-5} moles. Since the number of moles of $Na_2Cr_2O_7$ equals its molarity M times its volume V, then

$$3.1 \times 10^{-5} = 0.0020 \text{ V}$$
$$\text{so V} = 1.6 \times 10^{-2} \text{ L} = 16 \text{ mL}$$

Problem 10-9

In each case, the solubility is equal to the molarity of the iron ion since, for example,

$$Fe(OH)_2 \text{ (s)} \longrightarrow Fe^{2+} + 2\ OH^-$$

At pH = 8, pOH = 6

so $[OH^-] = 1 \times 10^{-6}$ M

For Fe $(OH)_2$,

$$K_{sp} = [Fe^{2+}]\ [OH^-]^2$$

so
$$[Fe^{2+}] = K_{sp} / [OH^-]^2 = 7.9 \times 10^{-15} / (10^{-6})^2 = 7.9 \times 10^{-3} \text{ M}$$

For Fe $(OH)_3$,

$$K_{sp} = [Fe^{3+}]\ [OH^-]^3$$

so
$$[Fe^{3+}] = K_{sp} / [OH^-]^3 = 6.3 \times 10^{-38} / (10^{-6})^3 = 6.3 \times 10^{-20} \text{ M}$$

To find the pH values at which the solubilities are 100 ppm, first convert this concentration into molarity:

100 ppm means 100 grams of iron in 1,000,000 grams of water

100 grams iron × 1 mol iron / 55.85 g iron = 1.79 mol Fe

1,000,000 g water = 1,000,000 mL water = 1000 L

Thus the molar concentration corresponding to 100 ppm is 1.79 mol / 1000 L = 1.79×10^{-3} M.

For $Fe(OH)_2$, we know $[Fe^{2+}]\ [OH^-]^2 = 7.9 \times 10^{-15}$,

so $[OH^-]^2 = 7.9 \times 10^{-15} / [Fe^{2+}] = 7.9 \times 10^{-15} / 1.79 \times 10^{-3} = 4.4 \times 10^{-12}$

so taking the square root of both sides, we find $[OH^-] = 2.1 \times 10^{-6}$, giving pOH = 5.68 and hence pH = 8.32.

Similarly, for $Fe(OH)_3$, we know $[Fe^{3+}][OH^-]^3 = 6.3 \times 10^{-38}$,

so $[OH^-]^3 = 6.3 \times 10^{-38} / [Fe^{3+}] = 6.3 \times 10^{-38} / 1.79 \times 10^{-3} = 3.52 \times 10^{-35}$,

so taking the cube root, $[OH^-] = 3.3 \times 10^{-12}$, giving pOH = 11.49, and hence pH = 2.51.

Problem 10-10

The pE equation given in the text for this equilibrium is

$$pE = 14.15 - (5/4)\,pH - (1/8)\log([NH_4^+]/[NO_3^-])$$

(a) Substituting pE = 11 and pH = 6, then

$$11 = 14.15 - (5/4) \times 6 - (1/8)\log([NH_4^+]/[NO_3^-])$$

so $\log([NH_4^+]/[NO_3^-]) = -8 \times (4.35) = -34.8$

and $[NH_4^+]/[NO_3^-] = 1.6 \times 10^{-35}$

(b) Substituting pE = –3 and pH = 6, then

$$-3 = 14.15 - (5/4) \times 6 - (1/8)\log([NH_4^+]/[NO_3^-])$$

so $\log([NH_4^+]/[NO_3^-]) = 8 \times (9.65) = 77.2$

and $[NH_4^+]/[NO_3^-] = 1.6 \times 10^{77}$

Problem 10-11

From the text, for the concentration of ions

$$pE = 13.2 + \log([Fe^{3+}]/[Fe^{2+}])$$

When the ion concentration ratio is 100:1, then

$$pE = 13.2 + \log(100) = 15.2$$

Problem 10-12

From the text, for the reaction involving CH_4 and CO_2

$$pE = 2.87 - pH + (1/8)\log(P_{CO_2}/P_{CH_4})$$

At pH = 4, we can solve for the desired ratio and obtain

$$\log(P_{CO_2}/P_{CH_4}) = (pE - 2.87 + 4) \times 8$$
$$= (pE + 1.13) \times 8$$

For the O_2 reaction, from the text

$$\begin{aligned} pE &= 20.75 - pH + (1/4)\log P_{O_2} \\ &= 20.75 - 4 + (1/4) \times \log(0.10) \\ &= +16.5 \end{aligned}$$

Thus $\log(P_{CO_2}/P_{CH_4}) = (16.5 + 1.13) \times 8 = 141$

and $P_{CO_2}/P_{CH_4} = 10^{141}$

Problem 10-13

(a) The unbalanced half-reaction is

$$NO_3^- \longrightarrow NO_2^-$$

Using standard redox balancing methods, we add H_2O to the right side to balance oxygen, 2 H^+ to the left to balance H, and 2 electrons to the left to balance the charge:

$$NO_3^- + 2\,H^+ + 2e^- \longrightarrow NO_2^- + H_2O$$

(b) Since by definition

$$\begin{aligned} pE &= E/0.0591, \text{ then} \\ pE &= +0.881/0.0591 = +14.9 \text{ here.} \end{aligned}$$

(c) From the balanced equation in (a), we can obtain the one-electron process:

$$\tfrac{1}{2}\,NO_3^- + H^+ + e^- \longrightarrow \tfrac{1}{2}\,NO_2^- + \tfrac{1}{2}\,H_2O$$

$$\begin{aligned} pE &= pE^\circ - \log([NO_2^-]^{1/2}/[NO_3^-]^{1/2}[H^+]) \\ &= pE^\circ + \log[H^+] - (\tfrac{1}{2})\log([NO_2^-]/[NO_3^-]) \\ &= pE^\circ - pH - (\tfrac{1}{2})\log([NO_2^-]/[NO_3^-]) \\ &= 14.9 - pH - (\tfrac{1}{2})\log([NO_2^-]/[NO_3^-]) \end{aligned}$$

(d) When $[NO_3^-]/[NO_2^-] = 100$, then by substitution

$$pE = 14.9 - pH - (\tfrac{1}{2})\log(1/100) = 14.9 - pH - (-2)/2$$

So $pE + pH = 15.9$ for this condition.

(e) Susbtituting $pE = 12$ and $pH = 5$ into the first equation for part (c), we obtain

$$12 = 14.9 - 5 - (\tfrac{1}{2})\log([NO_2^-]/[NO_3^-])$$

so $\log([NO_2^-]/[NO_3^-]) = -2.1 \times 2$

and $[NO_2^-]/[NO_3^-] = 10^{-4.2} = 6 \times 10^{-5}$.

Problem 10-14

When $CaSO_4$ dissolves in water, the process is

$$CaSO_4(s) \rightleftharpoons Ca^{2+} + SO_4^{2-}$$

Hence,

$$K_{sp} = [Ca^{2+}]\,[SO_4^{2-}]$$

If no reaction of sulfate with water occurs, then

$$[Ca^{2+}] = [SO_4^{2-}]$$

so $[SO_4^{2-}]^2 = 3.0 \times 10^{-5}$, because this is the value of K_{sp}.

Consequently,

$$[SO_4^{2-}] = 5.5 \times 10^{-3}$$

Thus, the solubility of $CaSO_4$ is 5.5×10^{-3} moles / L if no reaction with water occurs; because the molar mass of $CaSO_4$ is 136.14, the solubility corresponds to 0.75 grams per liter.

Problem 10-15

For reaction 6, we deduced that $K = K_{sp}K_b = K_{sp}K_w / K_a$.

Since at 5°C we are told

$$K_{sp} = 8.1 \times 10^{-9}$$
$$K_w = 0.2 \times 10^{-14}$$
$$K_a = 2.8 \times 10^{-11}$$

thus

$$K = (8.1 \times 10^{-9}) \times (0.2 \times 10^{-14}) / (2.8 \times 10^{-11})$$
$$= 5.8 \times 10^{-13}$$

Thus, because

$$S^3 = 5.8 \times 10^{-13}$$

we obtain, upon taking the cube root of both sides

$$S = 8.3 \times 10^{-5}\ M$$

Since this is less than that of 9.9×10^{-5} M calculated (see text) for 25°C, we conclude that the solubility of $CaCO_3$ increases with increasing temperature.

Problem 10-16

Reaction 3: $CO_2\,(g) + H_2O \rightleftharpoons H_2CO_3$

Reaction 1: $H_2CO_3 \rightleftharpoons H^+ + HCO_3^-$

Summing the reactions, and cancelling the common H_2CO_3 term, gives

$$CO_2\,(g) + H_2O \rightleftharpoons H^+ + HCO_3^-$$

Since this reaction is the sum of (1) and (3), its

$$K = K_1K_3 = (4.5 \times 10^{-7}) \times (3.4 \times 10^{-2}) = 1.5 \times 10^{-8}$$

Since by the Law of Mass Action

$$K = [H^+]\,[HCO_3^-] / P_{CO_2}$$

And since from the stoichiometry $[HCO_3^-] = [H^+]$,

we have $1.5 \times 10^{-8} = [H^+]^2 / 0.00039$

So $[H^+]^2 = 0.00039 \times 1.5 \times 10^{-8} = 5.85 \times 10^{-12}$

and thus $[H^+] = 2.42 \times 10^{-6}$, giving pH = 5.62

Problem 10-17

For the overall reaction, we know

$$K = K_{sp}\,K_b\,K_H\,K_1 / K_w$$

Here $K_b = 0.2 \times 10^{-14} / 2.8 \times 10^{-11}$
$= 7.1 \times 10^{-5}$

Substituting the 5°C values, we obtain

$$K = (8.1 \times 10^{-9})\,(7.1 \times 10^{-5})\,(0.065)\,(3.0 \times 10^{-7}) / (0.2 \times 10^{-14})$$
$$= 5.6 \times 10^{-6}$$

Equating K to $[Ca^{2+}]\,[HCO_3^-]^2 / P$

and substituting $P = 0.00039$
and using $S = [Ca^{2+}]$
so then $2S = [HCO_3^-]$

then $S(2\,S)^2 / 0.00039 = 5.6 \times 10^{-6}$

or $S^3 = 5.46 \times 10^{-10}$

and thus $S = 8.2 \times 10^{-4}$ M

Problem 10-18

The concentration of HCO_3^- will be determined by the equilibrium

$$HCO_3^- + H_2O \rightleftharpoons H_2CO_3 + OH^-$$

For which

$$K_b = K_w / K_1 = (1.0 \times 10^{-14}) / (4.5 \times 10^{-7}) = 2.2 \times 10^{-8} = [H_2CO_3]\,[OH^-] / [HCO_3^-]$$

Now from the solution to Problem 3-15, $[H_2CO_3] = 1.3 \times 10^{-5}$ M

so

$$\begin{aligned} [HCO_3^-] &= [H_2CO_3]\,[OH^-] / 2.2 \times 10^{-8} \\ &= 1.3 \times 10^{-5}\,[OH^-] / 2.2 \times 10^{-8} \\ &= 5.9 \times 10^{2}\,[OH^-] \end{aligned}$$

Now

$$\begin{aligned} &\text{at pH} = 6, \text{pOH} = 8, \text{and } [OH^-] = 10^{-8} \\ &\text{at pH} = 5, \text{pOH} = 9, \text{and } [OH^-] = 10^{-9} \\ &\text{at pH} = 4, \text{pOH} = 10, \text{and } [OH^-] = 10^{-10} \end{aligned}$$

and therefore

$$\begin{aligned} [HCO_3^-] &= 5.9 \times 10^{-6} \text{ M at pH} = 6 \\ &= 5.9 \times 10^{-7} \text{ M at pH} = 5 \\ &= 5.9 \times 10^{-8} \text{ M at pH} = 4 \end{aligned}$$

Problem 10-19

$$K_1 \text{ for } H_2CO_3 = \frac{[HCO_3^-][H^+]}{[H_2CO_3]}$$

so when $[H_2CO_3] = [HCO_3^-]$, then $[H^+] = K_1$ for H_2CO_3

$$= 4.5 \times 10^{-7}$$

i.e., pH = 6.3

$$\text{Similarly, } K_a \text{ for } HCO_3^- = \frac{[CO_3^{2-}][H^+]}{[HCO_3^-]} = K_2$$

so when $[CO_3^{2-}] = [HCO_3^-]$, then $[H^+] = K_2$

$$= 4.7 \times 10^{-11}$$

i.e., pH = 10.3

Problem 10-20

Since CaF_2 dissolves as per the reaction

$$CaF_2\ (s) \rightleftharpoons Ca^{2+} + 2\ F^-$$

thus, the solubility product $K_{sp} = [Ca^{2+}]\,[F^-]^2$

The product of ion concentrations will just reach the K_{sp} value before the salt begins to precipitate, so we can deduce the maximum fluoride concentration by substituting the known calcium value into it:

$$[F^-]^2 = K_{sp} / [Ca^{2+}]$$

$$= 4 \times 10^{-11} / 3.8 \times 10^{-4} = 1.05 \times 10^{-7}$$

Taking the square root, we obtain $[F^-] = 3.24 \times 10^{-4}$ M

To convert to the ppb scale, which is a mass / mass basis for solutions, we first convert moles of F to grams:

$$3.24 \times 10^{-4} \text{ moles F} \times 19.00 \text{ g F / mole F} = 6.16 \times 10^{-3} \text{ g F}$$

One liter of solution is 1000 grams because water (and a dilute solution) has a density of 1.0 g / mL, so the concentration of fluoride is 6.16×10^{-3} g F / 1000 g solution. Multiplying top and bottom of this ratio converts it to the basis of one million grams of solution, as required, so the concentration of fluoride is 6.16 grams F / 10^6 grams solution, that is, 6.16 ppm.

Problem 10-21

The ratio $[HCO_3^-] / [CO_3^{2-}]$ is determined by the equilibrium constant for the acid-base reaction that connects them,

$$K_b = [HCO_3^-]\,[OH^-] / [CO_3^{2-}] = K_5 \text{ of Table 10-3}$$

Rearranging, we obtain the requested ratio:

$$\frac{[HCO_3^-]}{[CO_3^{2-}]} = \frac{K_5}{[OH^-]} = \frac{2.1 \times 10^{-4}}{[OH^-]}$$

At pH = 4, pOH = 10, so $[OH^-] = 10^{-10}$ and the ratio is 2.1×10^6.
At pH = 8.5, pOH = 5.5, so $[OH^-] = 3.2 \times 10^{-6}$ and the ratio is 66.

Similarly, the ratio $[H_2CO_3] / [HCO_3^-]$ is determined by the equilibrium constant

$$K_a = [H^+]\,[HCO_3^-] / [H_2CO_3] = K_1 \text{ of Table 10-3}$$

$$\frac{[H_2CO_3]}{[HCO_3^-]} = \frac{[H^+]}{K_1} = \frac{[H^+]}{4.5 \times 10^{-7}}$$

At pH = 4, $[H^+]$ = 10^{-4}, so this ratio is 220.
At pH = 8.5, $[H^+]$ = 3.2×10^{-9}, so this ratio is 7.0×10^{-3}.

Considering both ratios simultaneously, then

$$\text{at pH} = 4 \quad [H_2CO_3] > [HCO_3^-] >> [CO_3^{2-}]$$
$$\text{and at pH} = 8.5 \quad [HCO_3^-] > [CO_3^{2-}] >> [H_2CO_3]$$

Thus, the statements made in the text are correct.

Problem 10-22

For water saturated with both CO_2 and $CaCO_3$, from Table 10-3 we have

$$[HCO_3^-] = 1.0 \times 10^{-3}\text{ M}$$
$$[OH^-] = 1.8 \times 10^{-6}\text{ M}$$
$$[H^+] = 5.6 \times 10^{-9}\text{ M}$$

and $[CO_3^{2-}] = 8.8 \times 10^{-6}$ M

The phenolphthalein alkalinity = $[CO_3^{2-}] = 8.8 \times 10^{-6}$ M

and total alkalinity $= 2\,[CO_3^{2-}] + [HCO_3^-] + [OH^-] - [H^+]$
$= 2 \times 8.8 \times 10^{-6} + 1.0 \times 10^{-3} + 1.8 \times 10^{-6} - 5.6 \times 10^{-9}$
$= 1.02 \times 10^{-3}$ M

Problem 10-23

Since $[CO_3^{2-}]$ = phenolphthalein alkalinity,

then $[CO_3^{2-}] = 3.5 \times 10^{-5}$ M here.

Since pH = 10, pOH = 4 and so $[OH^-] = 1.0 \times 10^{-4}$ M

Total alkalinity $= 2\,[CO_3^{2-}] + [HCO_3^-] + [OH^-] - [H^+]$
$= 2 \times 3 \times 10^{-5} + 1 \times 10^{-4} + 1 \times 10^{-4} - 10^{-10}$
$= 2.6 \times 10^{-4}$ M

Problem 10-24

Since the hardness index is based upon molar concentration, we convert the masses of the ions first into molar amounts

$$0.0040\text{ g Ca} \times \frac{1\text{ mole Ca}}{40.1\text{ g Ca}} = 0.00010\text{ moles Ca}$$

$$0.0012\text{ g Mg} \times \frac{1\text{ mole Mg}}{24.3\text{ g Mg}} = 0.000049\text{ moles Mg}$$

Thus since the volume is 0.5 L, the ion concentrations are

$$[Ca^{2+}] = 0.0001 / 0.5 = 0.0002 \text{ M, and } [Mg^{2+}] = 0.000049 / 0.5 = 0.00010 \text{ M.}$$

So Hardness = 0.00020 + 0.00010 = 0.00030 M = [$CaCO_3$ equivalent]

Thus Hardness = 0.00030 moles / L × 100.1 g / mole = 0.030 g $CaCO_3$ / L

= 30 mg $CaCO_3$ / L

Problem 10-25

From the calculations quoted in Table 10-4,

$$[Ca^{2+}] = 5.3 \times 10^{-4} \text{ M}$$

so hardness = 5.3×10^{-4} moles Ca^{2+} / L

This is equivalent to 5.3×10^{-4} moles $CaCO_3$ per liter, and because the molar mass of $CaCO_3$ is 100.0 g / mole, the hardness is 5.3×10^{-2} g $CaCO_3$ / L = 53 mg $CaCO_3$ / L. Thus, the U.S. median hardness value is a little less than that calculated under the assumption that equilibrium exists between $CaCO_3$(s) and CO_2(g).

Problem 10-26

If pH = 5.5, then pOH = 14 − 5.5 = 8.5
and hence $[OH^-] = 3.2 \times 10^{-9}$ M

From the text, we know

$$[Al^{3+}][OH^-]^3 = 1.0 \times 10^{-33}$$

so $[Al^{3+}] = 1.0 \times 10^{-33} / (3.2 \times 10^{-9})^3$
$= 3.1 \times 10^{-8}$ M

Since the molar mass of Al is 26.98 grams, the mass per liter is 8.2×10^{-7} g.

Problem 10-27

The relevant equilibrium here is

$$Al(OH)_3(s) \rightleftharpoons Al^{3+} + 3\ OH^-$$

for which

$$[Al^{3+}][OH^-]^3 = K_{sp} = 1.0 \times 10^{-33}$$

For $[Al^{3+}] = 0.020$, then

$$[OH^-]^3 = 1.0 \times 10^{-33} / 0.020$$

so $[OH^-] = 3.7 \times 10^{-11}$

i.e., $pOH = 10.4$, so $pH = 3.6$

Green Chemistry Problems

1. Scouring removes the outermost layer of the raw cotton fiber, known as the cuticle, which is composed of fats, waxes, and pectin. The fats and waxes make the fiber waterproof and thus in order to make the fiber wetable (required for subsequent steps of bleaching and dyeing), the cuticle must be removed.

2. (a) The use of sodium hydroxide solutions at elevated temperatures:

   - requires neutralization with acetic acid and concomitant aqueous wastes
   - requires large amounts of water for rinsing
   - is not selective, thus destroying not only the target fats and waxes but additional organic materials (including cellulose), producing aqueous waste with high BOD and COD
   - requires energy for the elevated temperature of the hydroxide solution

   Sodium hydroxide is a strong base and is very caustic.

   (b) Biopreparation:

   - eliminates neutralization with acetic acid and concomitant aqueous wastes
   - eliminates large amounts of water for rinsing
   - is very selective, thus destroying only the target pectin polymer, producing aqueous waste with BOD lowered by 20% and COD lowered by 50%
   - reduces energy requirements since this process is done at ambient temperature
   - eliminates the use of sodium hydroxide

3. (a) 2. Alternative reaction conditions for green chemistry.

   (b) 1. Prevention of waste.

   3. Use of substances that possess little or no toxicity to human health and the environment.

   4. Preserving efficacy of function while reducing toxicity.

   6. Reduction of energy requirements … performed at ambient temperatures.

Additional Problems

2. From the properties of the CO_2 sample, we can deduce the number of moles of it using the ideal gas law

$$PV = nRT$$

$$\text{So } n = PV/RT = 0.96 \times (0.025) / 0.082 \times (273 + 22)$$
$$= 0.00099 \text{ moles } CO_2$$

Thus the moles of carbon in the original sample must also have been 0.00099, which has a mass of 0.012 grams because carbon's molar mass is 12. Thus the concentration of carbon in the sample is 0.012 g/5000 g (because 5.0 L of water has a mass close to 5000 grams), or 2.4×10^{-6}, which is equivalent to 2.4 ppm.

The molar concentration of carbon, and hence of CH_2O, in the original sample is (0.00099 moles/5.0 L) = 0.00020 M. The complete oxidation of the organic matter to CO_2 is given by the balanced equation

$$CH_2O + O_2 \longrightarrow CO_2 + H_2O$$

Thus the moles of O_2 = moles of CH_2O, and hence the molar concentration of oxygen that would be consumed is also 0.00020 M. We can convert to mass of O_2 via

$$0.00020 \text{ moles } O_2 \times \frac{32.0 \text{ g } O_2}{1 \text{ mole } O_2} = 0.0063 \text{ g } O_2$$

Thus the oxygen demand is 6.3 milligrams of O_2 per liter.

3. (a) For the reaction

$$SO_4^{2-} \longrightarrow H_2S \text{ (acid)}$$

we balance oxygen (by adding H_2O), hydrogen (by adding H^+), and charge (by adding electrons) to obtain

$$SO_4^{2-} + 10\,H^+ + 8e^- \longrightarrow H_2S + 4\,H_2O$$

(b) The balanced one-electron process is

$$1/8\,SO_4^{2-} + 5/4\,H^+ + e^- \longrightarrow 1/8\,H_2S + 1/2\,H_2O$$

Thus

$$pE = pE° - \log (P_{H_2S}{}^{1/8} / [SO_4^{2-}]^{1/8} [H^+]^{5/4})$$
$$= pE° - (5/4)\,pH - (1/8) \log (P_{H_2S} / [SO_4^{2-}])$$
$$= 5.75 - (5/4)\,pH - (1/8) \log (P_{H_2S} / [SO_4^{2-}])$$

(c) By rearrangement of the equation in part (b),

$$-(1/8) \log (P_{H_2S} / [SO_4^{2-}]) = pE - pE° + 5/4\, pH$$

For any solution in equilibrium with atmospheric O_2, we know (see text) that

$$pE = 20.75 - pH + (1/4) \log P_{O_2}$$

and because $P_{O_2} = 0.21$ and $pH = 6$, here,

$$pE = 20.75 - 6 + (1/4) \log (0.21) = +14.58$$

We substitute this value into the equation for pE of the H_2S / SO_4^{2-} system above to obtain

$$-(1/8) \log (P_{H_2S} / [SO_4^{2-}]) = 14.58 - pE° + (5/4) \times 6$$
$$= 14.58 - 5.75 + 7.5$$
$$= 16.33$$
$$\text{or } \log (P_{H_2S} / [SO_4^{2-}]) = -131$$

Since $[SO_4^{2-}] = 10^{-5}$, then

since $P_{H_2S} / [SO_4^{2-}] = 10^{-131}$

$\therefore P_{H_2S} = 10^{-131} \times 10^{-5}$

$= 10^{-136}$ atm

Thus the partial pressure of H_2S under these conditions is absolutely negligible.

4. Since lead, like calcium, produces a 2+ ion, the reactions are identical to the calcium carbonate system:

$$PbCO_3\ (s) \rightleftharpoons Pb^{2+} + CO_3^{2-}$$
$$CO_3^{2-} + H_2O \rightleftharpoons HCO_3^- + OH^-$$

Thus, the overall reaction obtained by summing these together is

$$PbCO_3\ (s) + H_2O \rightleftharpoons Pb^{2+} + HCO_3^- + OH^-$$

for which

$$K = K_{sp} K_b = [Pb^{2+}] [HCO_3^-] [OH^-] = S^3$$

Now

$$K = 1.5 \times 10^{-13} \times 2.1 \times 10^{-4} = 3.15 \times 10^{-17} = S^3$$

which yields

$$S = 3.2 \times 10^{-6}\ M$$

If we assume instead that the reaction of carbonate with water is negligible here, then

$$K_{sp} = [Pb^{2+}] [CO_3^{2-}] = S^2 = 1.5 \times 10^{-13}$$

and hence

$$S = 3.9 \times 10^{-7}\ M$$

Yes, the results differ by an order of magnitude, depending upon whether or not the reaction of carbonate with water is included.

5. The reactions of HCO_3^- as an acid and as a base will be those that produce H^+ and OH^- respectively:

ACID: $HCO_3^- \rightleftharpoons H^+ + CO_3^{2-}$
BASE: $HCO_3^- + H_2O \rightleftharpoons OH^- + H_2CO_3$

For action as an acid, $K_a = 4.7 \times 10^{-11}$ from Table 10-3.
For action as a base, $K_b = K_w / K_1 (H_2CO_7)$
$= 10^{-14} / 4.5 \times 10^{-7} = 2.2 \times 10^{-8}$.

Since $K_b >> K_a$, its dominant action in water will be as a base.
For a 0.01 M solution of sodium bicarbonate, and hence for $[HCO_3^-]_0 = 0.01$, since

	$HCO_3^- + H_2O \rightleftharpoons$	OH^-	$+ H_2CO_3$
Initial	0.01	0	0
Equilibrium	0.01 – x	x	x

Thus $x^2 / (0.01 - x) = 2.2 \times 10^{-8}$
Assuming $0.01 >> x$, then $x^2 = 0.01 \times 2.2 \times 10^{-8}$
so $x = 1.5 \times 10^{-5}\ M = [OH^-] = [H_2CO_3]$
Thus pOH = 4.83 and hence pH = 9.17.

6. Now from its definition

hardness = $[Mg^{2+}] + [Ca^{2+}]$
in terms of $CaCO_3$ mass per liter

Thus a hardness of 30.0 mg/L can be converted to the molarity of $Ca^{2+} + Mg^{2+}$
Moles $CaCO_3 = 30.0 \times 10^{-3}\ g / 100.0\ g\ mole^{-1} = 3.0 \times 10^{-4}$ moles,
so $[Mg^{2+}] + [Ca^{2+}] = 3.0 \times 10^{-4}\ M$

But we are told $[Mg^{2+}] = 1.0 \times 10^{-4}\ M$,
so $[Ca^{2+}] = 3.0 \times 10^{-4} - 1.0 \times 10^{-4} = 2.0 \times 10^{-4}\ M$

We know by definition that

$[CO_3^{2-}]$ = phenolphthalein alkalinity = $1.0 \times 10^{-5}\ M$
and from the pH of 7.6
$\therefore\ [H^+] = 10^{-7.6} = 2.5 \times 10^{-8}\ M$
and thus $[OH^-] = 1 \times 10^{-14} / [H^+] = 1.0 \times 10^{-14} / 2.5 \times 10^{-8} = 4.0 \times 10^{-7}\ M$

Finally, the total alkalinity by definition is

$$2\,[CO_3^{2-}] + [HCO_3^-] + [OH^-] - [H^+]$$

so $6.2 \times 10^{-4} = 2 \times (1.0 \times 10^{-5}) + [HCO_3^-] + 4 \times 10^{-7} - 2.5 \times 10^{-8}$

so $[HCO_3^-] = 6.0 \times 10^{-4}\ M$

If the system is at equilibrium with respect to

$$HCO_3^- \rightleftharpoons H^+ + CO_3^{2-}$$

then the ratio $[H^+]\,[CO_3^{2-}]\,/\,[HCO_3^-]$ would equal K_a for HCO_3^-. Substituting the values above, we obtain for the ratio

$$2.5 \times 10^{-8} \times 1.0 \times 10^{-5} / 6.0 \times 10^{-4} = 4.2 \times 10^{-10}$$

Since $K_a = 4.7 \times 10^{-11}$, we conclude that the system is not quite at equilibrium.

If the system were to be saturated with $CaCO_3$, then $[Ca^{2+}]\,[CO_3^{2-}]$ would equal K_{sp} for $CaCO_3$. The actual product is $2 \times 10^{-4} \times 1.0 \times 10^{-5} = 2 \times 10^{-9}$, whereas K_{sp} is 4.6×10^{-9}, so the system is fairly close to equilibrium for this reaction.

7. From the stoichiometry of the set of reactions, we obtain:

$$\text{mol } O_2 = 0.5 \times \text{mol } Mn^{2+}$$

$$\text{mol } MnO_2 = \text{mol } Mn^{2+}$$

$$\text{mol } I_2 = \text{mol } MnO_2$$

$$\text{mol } Na_2S_2O_3 = 0.5 \times \text{mol } I_2$$

$$\therefore \text{mol } O_2 = \text{mol } Na_2S_2O_3/4$$

before incubation: mol O_2 = 10.15 mL × 1 L / 1000 mL × 0.00100 mol / L / 4 = 2.5×10^{-6}

concentration = 2.5×10^{-6} mol × 32.0 g / mol × 1000 mg / g / 0.0100 L = 8.0 mg/L

after incubation: mol O_2 = 2.40 mL × 1 L / 1000 mL × 0.00100 mol / L / 4 = 6.0×10^{-7}

concentration = 6.0×10^{-7} mol × 32.0 g / mol × 1000 mg / g / 0.0100 L = 1.9 mg/L

$\therefore$ BOD = 8.0 mg/L – 1.9 mg / L = 6.1 mg / L

This would be considered to be polluted, as it is well above the median BOD of 0.7 mg/l for unpolluted surface water in the United States.

CHAPTER 11

The Pollution and Purification of Water

Problem 11-1

Calcium hydroxide releases hydroxide ions:

$$Ca(OH)_2 \longrightarrow Ca^{2+} + 2\,OH^-$$

In solution, calcium bicarbonate exists as Ca^{2+} and HCO_3^- ions:

$$Ca(HCO_3)_2 \longrightarrow Ca^{2+} + 2\,HCO_3^-$$

In the presence of the base OH^-, HCO_3^- acts as an acid, donating H^+ to the base.

The ionic reaction is:

$$HCO_3^- + OH^- \longrightarrow H_2O + 2\,CO_3^{2-}$$

and so there must be a 1:1 ratio of OH^- to HCO_3^-.

Thus, the overall balanced reaction is

$$Ca(HCO_3)_2 + Ca(OH)_2 \longrightarrow 2\,CaCO_3\,(s) + 2\,H_2O$$

Thus, one mole of $Ca(OH)_2$ should be added for every mole of dissolved Ca^{2+}.

Problem 11-2

(a) From the diameter, we can calculate the volume of the particle, since

$$V = 4\pi r^3 / 3 = 4 \times 3.14 \times (0.5 \times 10^{-6}\,m)^3 / 3$$
$$= 5.2 \times 10^{-19}\,m^3$$

where the radius r is 0.5 μm here.

Since we are told the density is similar to that of 1 g / cm^3 for water, and since there are 10^6 cm^3 in one cubic meter, we can conclude that the mass of a particle is $10^6 \times 5.2 \times 10^{-19}$ or 5.2×10^{-13} grams.

To deduce the number of atoms present, we first need to find the mass in grams of one atom. Since

1 mole weighs 10 grams
so 6.02×10^{23} atoms weigh 10 grams
or 1 atom weighs 1.66×10^{-23} grams.

Thus the number of atoms present is

$$5.2 \times 10^{-13}\ \text{g} / (1.66 \times 10^{-23}\ \text{g atom}^{-1})$$
$$= 3 \times 10^{10}\ \text{atoms}$$

(b) Repeating the calculation for a 0.005 μm radius, we calculate 3×10^4 atoms.

Problem 11-3

Since N is –3 and as usual H is +1, the numbers for NH_2 add up to –1, so chlorine is +1 here. Since unlike charges attract, the negative ion from water—that is, OH^-—will attract and therefore extract the chlorine, giving HOCl, and leaving the H^+ to connect with the NH_2^- unit to give ammonia, NH_3:

$$NH_2Cl + H_2O \longrightarrow NH_3 + HOCl$$

Problem 11-4

First we calculate the mass of nitrate in the aquifer:

$$\frac{20\ \text{g nitrate}}{10^6\ \text{g water}} \times 10 \times 10^6\ \text{L water} \times \frac{1000\ \text{g water}}{1\ \text{L water}}$$
$$= 2.0 \times 10^5\ \text{grams nitrate}$$

We can convert the NO_3^- mass to moles, and thus to moles and mass of NH_3, because one mole of ammonia produces one mole of nitrate:

$$2.0 \times 10^5\ \text{g}\ NO_3^- \times \frac{1\ \text{mole}\ NO_3^-}{62.01\ \text{g}\ NO_3^-} \times \frac{1\ \text{mole}\ NH_3}{1\ \text{mole}\ NO_3^-} \times \frac{17.05\ \text{g}\ NH_3}{1\ \text{mole}\ NH_3}$$
$$= 5.5 \times 10^4\ \text{grams}\ NH_3$$

Problem 11-5

The molar masses of N and NO_3 are 14.0 and 62.0 grams respectively, so the fraction of nitrate that is nitrogen is 14.0 / 62.0 = 0.226. Thus a 50 ppm nitrate standard is equivalent to 0.226 × 50 = 11 ppm nitrogen standard. Thus the EU standard is slightly less stringent than that of the United States.

Problem 11-6

(i) We begin by identifying the oxidation numbers for the nitrogen atoms in the two ions.

Since H = +1, that for N in NH_4^+ must be –3 if the sum of the O.N's are to add up to the net ion charge of +1.

Since O = –2, that for N in NO_3^- must be +5 if the sum of the O.N's are to add up to the net ion charge of –1.

Thus the change in O.N. in the half-reaction is from –3 to +5, a loss of 8 e^-.

$$NH_4^+ \longrightarrow NO_3^- + 8\,e^-$$

We now balance for charge using H^+, so we must add 10 H^+ to the right-hand side:

$$NH_4^+ \longrightarrow NO_3^- + 8\,e^- + 10\,H^+$$

Finally, we balance for O, so add 3 H_2O to the left-hand side:

$$NH_4^+ + 3\,H_2O \longrightarrow NO_3^- + 8\,e^- + 10\,H^+$$

(ii) We identify the O.N. of the N in NO_2^- as +3, and that in N_2 as 0.

Balancing the N's, we see that there must be a six-electron gain in the process:

$$2\,NO_2^- + 6\,e^- \longrightarrow N_2$$

To balance charge, we add 8 H^+ to the left-hand side:

$$2\,NO_2^- + 6\,e^- + 8\,H^+ \longrightarrow N_2$$

To balance O, we add 4 H_2O to the right-hand side:

$$2\,NO_2^- + 6\,e^- + 8\,H^+ \longrightarrow N_2 + 4\,H_2O$$

Problem 11-7

The half-reaction for iron dissolution is

$$Fe\,(s) \longrightarrow Fe^{2+} + 2e^-$$

The reduction of water would yield H_2 (g)

$$H_2O \longrightarrow H_2\,(g)$$

In alkaline solution, this half-reaction can be balanced to

$$2\,H_2O + 2e^- \longrightarrow H_2 + 2\,OH^-$$

The overall net reaction is

$$Fe\ (s) + 2\,H_2O \longrightarrow Fe^{2+} + H_2\ (g) + 2\,OH^-$$

Problem 11-8

The unbalanced equation is

$$OH^- + Ca(HCO_3)_2 \longrightarrow CaCO_3\ (s) + ?$$

Since there is an excess of hydroxide ion, it will convert all the bicarbonate ion (the acid here) to carbonate ion:

$$2\,OH^- + 2\,HCO_3^- \longrightarrow 2\,CO_3^{2-} + 2\,H_2O$$

Since there is only one Ca^{2+} ion per two bicarbonates, only one carbonate will bind with calcium, leaving the other as a reaction product:

$$Ca^{2+} + CO_3^{2-} \longrightarrow CaCO_3\ (s)$$

Thus, the overall equation is

$$2\,OH^- + Ca(HCO_3)_2 \longrightarrow CaCO_3\ (s) + CO_3^{2-} + 2\,H_2O$$

The reaction products other than $CaCO_3$ are water and carbonate ion.

Problem 11-9

It would be unacceptable if the vinyl chloride were to remain unreacted, since it is strongly carcinogenic.

Problem 11-10

The half-reaction involving iron is

$$Fe \longrightarrow Fe^{2+} + 2e^-$$

The unbalanced half-reaction converting perchloroethene to ethene is

$$C_2Cl_4 \longrightarrow C_2H_4 + Cl^-$$

Balancing the (alkaline phase) half-reaction gives

$$C_2Cl_4 + 4\,H_2O + 8e^- \longrightarrow C_2H_4 + 4\,Cl^- + 4\,OH^-$$

The combined reaction involving $8e^-$ in both half-reactions is

$$4\,Fe + C_2Cl_4 + 4\,H_2O \longrightarrow 4\,Fe^{2+} + C_2H_4 + 4\,Cl^- + 4\,OH^-$$

Problem 11-11

From the text and from the answer to Problem 11-10, we see that one mole of iron is required for each mole of chlorine, so 3 moles of Fe are required for one mole of C_2HCl_3 and four for each mole of C_2Cl_4.

We first determine for the specified volume (and hence mass) of groundwater the number of moles of TCE and of PCE present:

$$1000 \text{ g solution} \times \frac{270 \text{ g TCE}}{10^6 \text{g solution}} \times \frac{1 \text{ mole TCE}}{131.4 \text{ g TCE}}$$

$$= 0.00205 \text{ moles TCE} \times \frac{3 \text{ moles Fe}}{1 \text{ mole TCE}}$$

$$= 0.00615 \text{ moles Fe}$$

$$1000 \text{ g solution} \times \frac{53 \text{ g PCE}}{10^6 \text{ g solution}} \times \frac{1 \text{ mole PCE}}{165.8 \text{ g PCE}}$$

$$= 0.00032 \text{ moles PCE} \times \frac{4 \text{ moles Fe}}{1 \text{ mole PCE}}$$

$$= 0.00128 \text{ moles Fe}$$

Thus the total number of moles of Fe required is 0.00743, which corresponds to a mass of $0.00743 \times 55.85 = 0.42$ grams of Fe.

Problem 11-12

The reaction between ammonia and water is

$$NH_3 + H_2O \rightleftharpoons NH_4^+ + OH^-$$

so $K_b = [NH_4^+]\,[OH^-] / [NH_3]$

Solving this equation for the ratio of ammonia to ammonium ion concentrations gives

$$\begin{aligned}[NH_3] / [NH_4^+] &= [OH^-] / K_b \\ &= [OH^- / 1.8 \times 10^{-5} \\ &= 5.6 \times 10^4\,[OH^-]\end{aligned}$$

By taking logarithms of both sides, we can relate the ratio to the pOH and thus to pH:

$$\begin{aligned}\log([NH_3] / [NH_4^+]) &= \log(5.6 \times 10^4) + \log[OH^-] \\ &= 4.74 - \text{pOH}\end{aligned}$$

But pH + pOH = 14, so

$$\log([NH_3]/[NH_4^+]) = 4.74 - (14 - pH)$$
$$= pH - 9.26$$

For pH = 5, the log ratio = −4.26, so the ratio is 5.5×10^{-5}
For pH = 7, the log ratio = −2.26, so the ratio is 0.0055
For pH = 9, the log ratio = −0.26, so the ratio is 0.55
For pH = 11, the log ratio = +1.74, so the ratio is 55

Problem 11-13

The energy cost in boiling or freezing the water in huge amounts would be prohibitive.

Problem 11-14

The unbalanced half-reaction would be $CN^- \longrightarrow HCO_3^- + N_2$

Balancing the N requires 2 CN^-, which in turn gives two bicarbonate ions, so the half-equation balanced in elements other than H and O is:

$$2\,CN^- \longrightarrow 2\,HCO_3^- + N_2$$

From the text, we know the oxidation number here of C in CN^- is +2, and because H is always +1 and O is −2, that of the carbon in bicarbonate ion must be +4. The nitrogen in N_2 must have a zero oxidation number because it is in an uncharged elemental form.

Thus the sum of oxidation numbers on the left side of the latter equation is $2 \times (+2 -3) = -2$. The only elements on the right side whose oxidation numbers will change are C and N; their sum in the above equation is $2 \times (+4) + 2 \times (0) = +8$. Thus the change in oxidation numbers in the process is a 10-electron loss to the right side. We add $10e^-$ to the right side to balance it:

$$2\,CN^- \longrightarrow 2\,HCO_3^- + N_2 + 10e^-$$

Now the left side has a net electrical charge of $2 \times (-1) = -2$, and the right side $2 \times (-1) - 10 = -10$, giving a difference of −10 in excess on the right. Hence we must add 10 H^+ to it to balance charge:

$$2\,CN^- \longrightarrow 2\,HCO_3^- + N_2 + 10e^- + 10H^+$$

Finally, to balance oxygen, we note that the right side has 6 more O atoms than the left, so we add 6 H_2O to the left:

$$2\,CN^- + 6\,H_2O \longrightarrow 2\,HCO_3^- + N_2 + 10e^- + 10H^+$$

This half-reaction now is balanced.

Problem 11-15

Common species of nitrogen containing the element in more oxidized form than N_2 (i.e., with oxidation number > 0) are NO and NO_2 gases, and NO_2^- and NO_3^- ions, as listed in Table 10-2.

Problem 11-16

Since the acid reaction is

$$HCN \rightleftharpoons H^+ + CN^-$$

then $K_a = [H^+][CN^-]/[HCN]$
so $[CN^-] = K_a[HCN]/[H^+]$

The ratio of concern is that for CN^- divided by the total for $CN^- + HCN$.

By substitution for $[CN^-]$, then

$$\frac{[CN^-]}{[HCN] + [CN^-]} = \frac{K_a[HCN][H^+]^{-1}}{[HCN] + K_a[HCN][H^+]^{-1}} = \frac{K_a[H^+]^{-1}}{1 + K_a[H^+]}$$

Multiplying top and bottom by $[H^+]$, we obtain the expression $K_a/([H^+] + K_a)$ for the fraction desired. Since $K_a = 6 \times 10^{-10}$, then

$$\text{at pH} = 4, [H^+] = 10^{-4} \text{ so fraction} = 6 \times 10^{-10}/(10^{-4} + 6 \times 10^{-10})$$
$$= 6 \times 10^{-6}$$

Similarly, at pH = 7, fraction = 6×10^{-3}
and at pH = 10, fraction = 0.86

Problem 11-17

For the reaction

$$H_2O_2 \longrightarrow 2\,OH$$

then $\Delta H = 2\Delta H_f(OH) - \Delta H_f(H_2O_2)$
$= 2 \times (+39.0) - (-136.3)$
$= +214.3 \text{ kJ mol}^{-1}$

From Chapter 1,

$$\lambda = 119{,}627/E$$
$$= 119{,}627/214.3$$
$$= 558 \text{ nm}$$

For photons of λ = 254 nm, their energy is

$$E = 119{,}627 / \lambda$$
$$= 119{,}627 / 254$$
$$= 471.0 \text{ kJ mol}^{-1}$$

Thus the fraction of the photon's energy that is actually used is (214.3 / 471.0) = 0.45, or 45%.

Problem 11-18

The unbalanced reaction is

$$S_2O_8^{2-} \longrightarrow SO_4^{2-}$$

This is easily balanced to give

$$S_2O_8^{2-} + 2e^- \longrightarrow 2\,SO_4^{2-}$$

The other half-reaction is

$$C_2H_2O_4 \longrightarrow CO_2$$

Using standard methods, the balanced version is

$$C_2H_2O_4 \longrightarrow 2\,CO_2 + 2\,H^+ + 2e^-$$

Thus the overall balanced reaction is

$$S_2O_8^{2-} + C_2H_2O_4 \longrightarrow 2\,CO_2 + 2\,SO_4^{2-} + 2\,H^+$$

In the titration calculation, we first convert the mass of oxalic acid into moles of it, then to moles of $S_2O_8^{2-}$, and then to the volume of its solution:

$$1000 \text{ g } C_2H_2O_4 \times \frac{1 \text{ mole } C_2H_2O_4}{90.04 \text{ g } C_2H_2O_4} \times \frac{1 \text{ mole } S_2O_8^{2-}}{1 \text{ mole } C_2H_2O_4}$$

$$= 11.1 \text{ moles } S_2O_8^{2-} \times \frac{1 \text{ L solution}}{0.010 \text{ moles } S_2O_8^{2-}}$$

$$= 1.1 \times 10^3 \text{ L}$$

Box 11-4, Problem 1

(a) [B] reaches its peak when d [B] / dt = 0. When we differentiate the expression given for [B] with respect to time, we obtain

$$\frac{d\,[B]}{dt} = k_1[A]_0(-k_1e^{-k_1t} + k_2e^{-k_2t})/(k_2 - k_1)$$

The term in parentheses in the numerator must be zero if the expression on the right side is to equal zero, so

$$k_2 e^{-k_2 t} = k_1 e^{-k_1 t}$$

or $$e^{t(k_1 - k_2)} = (k_1 / k_2)$$

Taking the natural logarithm of both sides, we obtain

$$t\,(k_1 - k_2) = \ln\,(k_1 / k_2)$$

so the time at which [B] peaks is

$$t = \frac{\ln\,(k_1 / k_2)}{k_1 - k_2}$$

Green Chemistry Problems

1. Iminodisuccinate is biodegradable and as a result places less of a load on the environment (see phosphates and how they act as nutrients), and it does not have to be removed in wastewater treatment plants in contrast to many chelating agents such as phosphates. The synthesis of iminodisuccinate from maleic anhydride is performed under mild conditions, uses only water as a solvent, and excess ammonia is recycled back into production of more iminodisuccinate.

2. (a) The design chemicals that are less toxic than current alternatives or are inherently safer with regard to accident potential.

   (b) 2. Synthetic methods should be designed to maximize the incorporation of all materials used in the process into the final product.

   3. Use of substances that possess little or no toxicity to human health and the environment.

   4. Preserving efficacy of function while reducing toxicity.

   5. The use of auxiliary substances (e.g., solvents, separation agents, etc.) should be made unnecessary whenever possible and innocuous when used.

   10. Chemical products should be designed so that at the end of their function they do not persist in the environment and break down into innocuous degradation products.

Additional Problems

1. The equilibrium is $HOCl \rightleftharpoons H^+ + OCl^-$, where $[H^+]$ is determined by other species.

   $$\text{Since } K_a = \frac{[H^+][OCl^-]}{[HOCl]}$$

   the equilibrium concentration of un-ionized (i.e., molecular) acid is expressed, using brackets in the usual way, as [HOCl], whereas the original concentration of HOCl (before ionization) can be expressed (as "$[HOCl]_0$") in terms of equilibrium concentrations of ionized species in the following way:

   $$[HOCl]_0 = [HOCl] + [OCl^-]$$

   Here the subscript zero denotes the original concentration.

   $$\text{Thus, the fraction un-ionized} = [HOCl] / [HOCl]_0 = \frac{[HOCl]}{[OCl^-] + [HOCl]}$$

   From the K_a expression, $[OCl^-] = K_a [HOCl] / [H^+]$

   Substituting for $[OCl^-]$ in the "fraction" expression, we obtain

   $$\text{un-ionized fraction} = \frac{[HOCl]}{[HOCl](1 + K_a/[H^+])} = \frac{1}{1 + K_a/[H^+]}$$

   $$= \frac{[H^+]}{K_a + [H^+]} = \frac{[H^+]}{2.7 \times 10^{-8} + [H^+]}$$

At pH = 7.0	$[H^+] = 1.00 \times 10^{-7}$	fraction = 0.79
7.5	$= 3.16 \times 10^{-8}$	= 0.54
8.0	$= 1.00 \times 10^{-8}$	= 0.27
8.5	$= 3.16 \times 10^{-9}$	= 0.10

   At pH = 8.5 (or even 8.0), the majority of the chlorine would be in the relatively ineffective OCl^- form, so it would not be a good idea.

2. First convert mg/L to mol/L : 0.5 mg / L × 0.001 g / mg / 70.91 g / mol = 7.05×10^{-6} mol/L

   Then using $K_H = [Cl_2] / P_{Cl_2}$:

   $$P_{Cl_2} = [Cl_2] / K_H = 7.05 \times 10^{-6}\ \text{M} / 8.0 \times 10^{-3}\ \text{M atm}^{-1} = 8.8 \times 10^{-4}\ \text{atm}$$

3. Set up an equilibrium table:

	$HOCl\ (aq)$	$\rightleftharpoons$	$H^+\ (aq)$	$+\ OCl^-\ (aq)$	$K_a = 3.5 \times 10^{-8}$
initial:	1		0	0	
change:	$-x$		x	x	
equilibrium:	$1 - x$		x	x	

$K_a = [H^+]\,[OCl^-] / [HOCl]$

$\therefore 3.5 \times 10^{-8} = x^2 / (1.00 - x) \approx x^2$ (assumes $x << 1.00$)

$\therefore x = \sqrt{3.5 \times 10^{-8}} = 1.87 \times 10^{-4}$

$\therefore [H^+] = [OCl^-] = 1.87 \times 10^{-4}$ M; $[HOCl] = 1.00\text{ M} - 1.87 \times 10^{-4}\text{ M} = 1.00$ M

$pH = -\log(1.87 \times 10^{-4}) = 3.73$

percent dissociated $= [H^+] / [HOCl]_o \times 100\% = 1.87 \times 10^{-4}\text{ M}/1.00\text{ M} \times 100\% = 0.019\%$

Similarly for 0.100 M: $x = \sqrt{(0.100 \times 3.5 \times 10^{-8})} = 5.92 \times 10^{-5}$

$[H^+] = [OCl^-] = 5.92 \times 10^{-5}$ M; $pH = 4.23$

percent dissociated $= 5.92 \times 10^{-5}\text{ M} / 0.100\text{ M} \times 100\% = 0.059\%$

4. The reaction is

$$Cl_2 + H_2O \rightleftharpoons H^+ + Cl^- + HOCl$$

so $K = [H^+]\,[Cl^-]\,[HOCl] / [Cl_2] = 4.5 \times 10^{-4}$

Let the concentration of Cl_2 that reacts be y, so

$[Cl^-] = [HOCl] = y$
and $[Cl_2] = [Cl_2]_0 - y$

The molar initial concentration of Cl_2 must first be obtained:

$$\frac{50\text{ g } Cl_2}{10^6\text{ g } H_2O} \times \frac{1000\text{ g } H_2O}{1\text{ L } H_2O} \times \frac{1\text{ mole } Cl_2}{70.9\text{ g } Cl_2} = 7.05 \times 10^{-4}\text{ M}$$

Thus we have

$[H^+]\,y^2 / (7.05 \times 10^{-4} - y) = 4.5 \times 10^{-4}$

or $y^2\,[H^+] = 3.17 \times 10^{-7} - 4.5 \times 10^{-4}\,y$

or $y^2\,[H^+] + 4.5 \times 10^{-4}y - 3.17 \times 10^{-7} = 0$

For pH = 0, $[H^+] = 1.0$ M, and solving the quadratic equation gives—as the only positive root—the value for $y = 3.81 \times 10^{-4}$, so the fraction reacted is $(3.81 \times 10^{-4} / 7.05 \times 10^{-4})$ or 0.54, so 46% remains unreacted. Similarly, at pH = 1, $[H^+] = 0.10$ M, which leads to y = 0.88, so 12% is unreacted. For pH = 2, only 1% is unreacted.

5. (a) By oxidation number rules, the H is +1 and the O is −2 in such compounds. Thus in HOCl, the sum of H and O is −1, so the Cl must be +1 for the sum of all numbers to equal zero. Similarly, in Cl_2O, because O is −2, each Cl must be +1. Since no oxidation numbers change during the reaction, we conclude that it is *not* a redox process.

   (b) By the Law of Mass Action, the equilibrium constant expression for the reaction must be:

   $$K = [Cl_2O] / [HOCl]^2$$

   We know $K = 9 \times 10^{-3}$, and that $[HOCl] = 3 \times 10^{-5}$,

   So upon rearrangement of the equation, we can solve for $[Cl_2O]$:

   $$[Cl_2O] = K\,[HOCl]^2 = 9 \times 10^{-3} \times (3 \times 10^{-5})^2 = 8 \times 10^{-12}$$

   Thus the equilibrium concentration of Cl_2O is very small, 8×10^{-12} M.

6. The formulas for those substances are, respectively, Cl_2, HOCl, ClO_2, NH_2Cl, and NaCl. Since the oxidation numbers for H, O, and N are +1, −2, and −3 and that for Na is +1, then using the principle that the sum of the oxidation numbers must add to zero in neutral compounds, we obtain:

   0 for Cl in Cl_2
   +1 for Cl in HOCl
   +4 for Cl in ClO_2
   +1 for Cl in NH_2Cl
   −1 for Cl in NaCl

   Since −1 is the stablest oxidation number, all the others will try to acquire extra electrons and hence be oxidizing agents. Presumably the order of oxidizing ability is proportional to the difference in charge from −1, so

   $$ClO_2 \gg HOCl = NH_2Cl > Cl_2$$

7. The formula for sodium tripolyphosphate is $Na_5P_3O_{10}$, so its molar mass is 367.86 grams. Thus one gram of it is equal to 1.00 / 367.86 = 0.00272 moles. Since there are 3 moles of P both in it and in $Ca_5(PO_4)_3OH$, the number of moles of the latter produced by 1 gram of the former is also 0.00272 moles. Since the molar mass of $Ca_5(PO_4)_3OH$ is 502.32 grams, the mass of it that is produced is 0.00272 × 502.32 = 1.366 grams. We can obtain the volume of this mass by inverting the density:

   $$1.366 \text{ grams} \times 1 \text{ mL} / 3.1 \text{ grams} = 0.441 \text{ mL}$$

If we estimate the annual mass of detergent used by the family to be about 50 kilograms, then the phosphate mass would have been 25 kilograms. Since by the calculations above we know that 0.441 mL is required for the calcium salt that is produced by one gram of sodium polyphosphate, then for 25 kilograms the volume would be 25,000 g × 0.441 mL / g = 11 liters.

8. We need concentration of each salt in M (mol/L); note that $kg/m^3 = g/L$. Convert g to mol; multiply by number of ions dissociated:

$NaCl$: 28.014 g / L / 58.44 g / mol = 0.479 mol / L × 2 = 0.958 mol / L solute

$MgCl_2$: 3.812 g / L / 95.22 g / mol = 0.0400 mol / L × 3 = 0.120 mol / L solute

$MgSO_4$: 1.752 g /L / 120.38 g / mol = 0.0146 mol / L × 2 = 0.0291 mol / L solute

$CaSO_4$: 1.283 g /L / 136.15 g / mol = 0.00942 mol / L × 2 = 0.0188 mol / L solute

K_2SO_4: 0.816 g / L / 174.27 g / mol = 0.00468 mol / L × 3 = 0.0140 mol / L solute

total concentration of solute M = 1.140 mol / L

$\pi = MRT = 1.140\ mol / L \times 0.082\ L\ atm\ K^{-1}\ mol^{-1} \times 293\ K = 27.4\ atm$

A pressure greater than 27.4 atm must be applied to seawater to force it to undergo reverse osmosis.

9. The solvents could be combusted using catalytic oxidation to innocuous products. This would be less harmful to the environment than simply boiling them away into the air, but it is an expensive process. Of course distilling off the contaminants to produce reusable solvent would be the optimum solution.

10. Hydrogen peroxide contains an O—O bond, so presumably peroxydisulfate and peroxymonosulfate anions do as well. Since the structure of sulfate ion has four oxygens surrounding the sulfur, that of peroxymonosulfate ion does as well, with the additional oxygen being in a peroxy unit:

```
        O
        ||
HO — S — O — O⁻
        ||
        O
```

In H_2O_2, since the H atoms are +1 each, then each oxygen is –1 rather than its regular –2 in oxidation number. To acquire the extra electrons, H_2O_2 must therefore act as an oxidizing agent.

11. (a) The addition of calcium hydroxide, $Ca(OH)_2$, would precipitate the phosphate as $Ca_3(PO_4)_2$ and would abstract hydrogen ions from both bicarbonate ion and ammonium ion, resulting in a precipitate of $CaCO_3$ and dissolved ammonia, which could be purged from the water by aeration. The salt could be removed by ion exchange.

(b) Denitrifying bacteria could be used to reduce the nitrite to N_2 gas, and then aeration at a high pH could be used to oxidize the Fe^{2+} to Fe^{3+} which would then precipitate and could be filtered. Finally, the TCE could be removed by one of several methods such as ozone / H_2O_2.

(c) Glucose could be converted to carbon dioxide by sprinkling the water over aerobic bacteria. Then, by producing a high pH by adding a base, cadmium could be precipitated as its hydroxide and removed. Finally CCl_4 could be removed by a reductive chlorination method.

12.

Oxidation number	Formula of example	Name of example
−1	Cl^-	Chloride ion
0	Cl_2	Molecular chlorine
+1	HOCl (or OCl^-)	Hypochlorous acid
+2	ClO	Chlorine monoxide
+3	ClO_2^-	Chlorite ion
+4	ClO_2	Chlorine dioxide
+5	ClO_3^-	Chlorate ion
+6	ClO_3	Chlorine trioxide
+7	ClO_4^-	Perchlorate ion

CHAPTER 12

Toxic Heavy Metals

Problem 12-1

Both H^+ ions come from the same H_2S molecule to produce the MS product:

$$M^{2+} + H_2S \longrightarrow MS + 2\,H^+$$

According to the text, R—S—H reacts with M^{2+} to produce R—S—M—S—R by release of hydrogen ions; clearly two RSH units are required.

$$M^{2+} + 2\,R{-}S{-}H \longrightarrow R{-}S{-}M{-}S{-}R + 2\,H^+$$

Problem 12-2

The reaction by which HgS dissolves is

$$HgS(s) \rightleftharpoons Hg^{2+} + S^{2-}$$

Here, the solubility $y = [Hg^{2+}] = [S^{2-}]$, so upon substitution into the K_{sp} expression

$$K_{sp} = [Hg^{2+}]\,[S^{2-}]$$

we obtain

$$3.0 \times 10^{-53} = y^2$$

so $\quad y = 5.5 \times 10^{-27}$ M

Since there are 6.02×10^{23} ions per mole (Avogadro's constant), this solubility corresponds to

$$5.5 \times 10^{-27} \text{ mole L}^{-1} \times 6.02 \times 10^{23} \text{ ions mole}^{-1}$$
$$= 0.0033 \text{ ions L}^{-1}$$

Inverting this value, we find that one ion would occupy a volume of 300 liters.

Problem 12-3

The steady-state concentration is related to the half-life $t_{0.5}$, here 70 days, and to the input rate R by the formula (see Chapter 5)

$$C_{ss} = 1.45\, R\, t_{0.5}$$

$$\text{The daily input rate } R = 1000 \text{ g fish} \times \frac{0.5 \text{ g methylmercury day}^{-1}}{10^6 \text{ g fish}}$$
$$= 0.0005 \text{ g methylmercury day}^{-1}$$

Substituting, we obtain

$$C_{ss} = 1.45 \times 0.0005 \text{ g methylmercury day}^{-1} \times 70 \text{ days}$$
$$= 0.051 \text{ g methylmercury}$$

Problem 12-4

From the percent composition data, we can deduce the empirical formula of a compound. Since it contains 26.1% chlorine, it must contain 100 – 26.1 = 73.9% mercury. Calculate the moles of each element in 100 g (or any other arbitrary amount) of the compound:

$$\text{moles chlorine} = 26.1 \text{ g} / 35.45 \text{ g mole}^{-1} = 0.736 \text{ moles}$$
$$\text{moles mercury} = 73.9 \text{ g} / 200.6 \text{ g mole}^{-1} = 0.368 \text{ moles}$$

Since the ratio of moles Cl to moles Hg is 2.0 to 1, the empirical formula is $HgCl_2$, which is consistent with the molecular formula $HgCl_2$.

Problem 12-5

The concentration of 0.50 ppm Hg corresponds to 0.50 g Hg in 10^6 grams of fish; thus, if the fish's mass is 1000 g, its total mercury content is:

$$1000 \text{ g fish} \times \frac{0.50 \text{ g Hg}}{10^6 \text{ g fish}} = 0.0005 \text{ g Hg}$$
$$= 0.5 \text{ mg Hg}$$

Since the concentration is 0.5 g Hg per 10^6 g fish, we can multiply the mass of mercury by the inverse of the concentration to obtain the mass of fish:

$$100 \times 10^{-3} \text{ g Hg} \times \frac{10^6 \text{ g fish}}{0.5 \text{ g Hg}} = 2.0 \times 10^5 \text{ g fish}$$

Problem 12-6

Calculate the mass of 0.30 ppm MeHg fish that contains the RfD for a 60 kg woman:

$$60 \text{ kg woman} \times \frac{0.1 \times 10^{-6} \text{g MeHg}}{1 \text{ kg woman / day}} \times \frac{10^{6} \text{g fish}}{0.3 \text{ g MeHg}}$$

$$= 20 \text{ g fish / day}$$

Thus, per week, a woman can consume 140 grams of such fish.

Problem 12-7

Since we know the Pb charge is +2 and the formula shows 3 of them, the total positive charge on ions is 3 × (+2) = +6. Thus the charges from the anions must total –6. Since the carbonate charge is –2, there could potentially be zero or one or two or three of them, giving charge from them totalling 0, –2, –4, and –6 respectively. Thus the number of singly negative hydroxide ions required to have the negative charge sum to –6 would be 6, 4, 2, and 0, respectively. Thus the possible formulas are:

$Pb_3(OH)_6$

$Pb_3CO_3(OH)_4$

$Pb_3(CO_3)_2(OH)_2$

$Pb_3(CO_3)_3$, i.e., $PbCO_3$

Problem 12-8

For a liquid, 10 ppb means 10 grams of contaminant per 10^9 grams of water. The mass of 2 liters of water is close to 2000 g, since its density is close to one gram per milliliter. Thus, the mass of lead in the daily amount of water is:

$$2000 \text{ g water} \times \frac{10 \text{ g lead}}{10^{9} \text{g water}} = 2.0 \times 10^{-5} \text{grams of lead}$$

Problem 12-9

The corresponding reactions for HgS in an acidic environment are, with data from the text:

$$HgS(s) \rightleftharpoons Hg^{2+} + S^{2-} \qquad (K_{sp} = 3.0 \times 10^{-53})$$

$$S^{2-} + H^{+} \rightleftharpoons HS^{-} \qquad (K = 7.7 \times 10^{12})$$

$$HS^{-} + H^{+} \rightleftharpoons H_2S \qquad (K' = 1.0 \times 10^{7})$$

Summing the three reactions, we obtain

$$HgS(s) + 2\,H^+ \rightleftharpoons Hg^{2+} + H_2S(aq)$$

$$K_{overall} = K_{sp}KK'$$
$$= 2.3 \times 10^{-33}$$

Now $K_{overall} = [Hg^{2+}]\,[H_2S]\,/\,[H^+]^2$
and $[Hg^{2+}] = [H_2S]$ from the stoichiometry, so

$$[Hg^{2+}]^2 = 2.3 \times 10^{-33}\,[H^+]^2$$

or $[Hg^{2+}] = 4.8 \times 10^{-17}\,[H^+]$

Even at pH = 2, where $[H^+] = 10^{-2}$, $[Hg^{2+}] = 4.8 \times 10^{-15}$, so that the solubility of PbS never amounts to very much, but it does, in fact, increase with increasing exposure to acid.

Problem 12-10

Given that the density of blood is about 1 g/mL, so 1 mL of it has a mass of 1 gram, and vice-versa. Thus the volume of 10^9 grams of blood is 10^9 mL, or 10^6 L, or 10^7 deciliters because 10 deciliters = 1 L.

Thus 60 ppb = 60 g Pb/10^7 deciliters
= 6.0×10^{-6} g Pb/dL
= 6.0 micrograms of Pb per deciliter of blood

Since 60 g Pb is present in 10^6 L, and since 60 g Pb $= 60\text{ g} \times \dfrac{1\text{ mole Pb}}{207.2\text{ g Pb}} = 0.29$ moles Pb

the concentration of lead can also be expressed as 0.29 moles Pb per 10^6 L, or 0.29×10^{-6} moles Pb per liter, i.e., there are 0.29 micromoles of lead per liter.

Green Chemistry Problems

1. (a) 3. The design of chemicals that are less toxic than current alternatives or inherently safer with regard to accident potential.

   (b) 4. Preserving efficacy of function while reducing toxicity.

2. Yttrium oxide is twice as effective as lead oxide on a weight basis and is 1/120 as toxic. Eliminates the use of 1 million pounds of lead, and also 25,000 pounds of chromium and 50,000 pounds of nickel (in pretreatment baths) on an annual basis.

3. There is lower air pollution due to decreased solvent emissions, better corrosion protection due to better coverage of poorly accessible areas (less steel used), and reduced waste due to high transfer efficiency.

4. (a) 3. The design of chemicals that are less toxic than current alternatives or inherently safer with regard to accident potential.

   (b) 4. Preserving efficacy of function while reducing toxicity.

Additional Problems

1. Assuming the Hg vapor behaves ideally, we can apply the ideal gas law:

   $n / V = P / RT = 1.6 \times 10^{-6}$ atm / (0.082 L atm K^{-1} $mol^{-1} \times 293$ K) = 6.7×10^{-8} mol L^{-1}

   Convert this to mg m^{-3}:

   6.7×10^{-8} mol $L^{-1} \times 200.6$ g $mol^{-1} \times 10^{3}$ mg / g $\times$ 1000 L m^{-3} = 13 mg m^{-3}

   This is much larger than the threshold limit value of 0.05 mg m^{-3}. (Obviously such a laboratory, which hopefully does not in fact exist anywhere, could not be used, and all of the floorboards, walls, benches, and ceiling tiles would need to be removed and destroyed.)

2. The solubilization reaction for PbS is:

   $$PbS\ (s) \rightleftharpoons Pb^{2+} + S^{2-} \quad K_{sp}\ (PbS)$$

   and the reaction of sulfide ion with water is:

   $$S^{2-} + H_2O \rightleftharpoons HS^- + OH^- \quad K_b\ (S^{2-})$$

   Summing these two reactions, we obtain

   $$PbS\ (s) + H_2O \rightleftharpoons Pb^{2+} + HS^- + OH^-$$

   for which the equilibrium constant K is the product of those for the two reactions:

   $$K = K_{sp}\ K_b$$

   Now for S^{2-} the K_b value can be obtained from that for its conjugate acid, HS^-:

   $$K_b\ (S^{2-}) = K_w / K_a\ (HS^-) = 1.0 \times 10^{-14} / 1.3 \times 10^{-13} = 7.7 \times 10^{-2}$$

   Thus, because (from the text) K_{sp} (PbS) = 8.4×10^{-28}, then

   $$K = 8.4 \times 10^{-28} \times 7.7 \times 10^{-2} = 6.5 \times 10^{-29}$$

   In terms of concentrations, from the balanced equation for the net reaction, we have:

   $$K = [Pb^{2+}]\ [HS^-]\ [OH^-]$$

   because by convention neither the solid nor water appears in the expression.

From the stoichiometry of the net equation, the solubility Y of PbS is related to the ion concentrations by:

$$Y = [Pb^{2+}] = [HS^-] = [OH^-]$$

so by substitution, $K = Y^3$

Solving the equation $Y^3 = 6.5 \times 10^{-29}$, we obtain $Y = 4.0 \times 10^{-10}$ M as the solubility.

If no reaction of sulfide with water were to occur, then the solubility would be determined by the first equation alone, for which $K_{sp} = [Pb^{2+}]\,[S^{2-}] = 8.4 \times 10^{-28}$. Here $[Pb^{2+}] = [S^{2-}]$, so the solubility $= [Pb^{2+}] = (8.4 \times 10^{-28})^{1/2} = 2.9 \times 10^{-14}$ M in this case.

3. (a) The general form for an exponential decay curve here is

$$P = K' e^{-aB}$$

where P is the percent of children having blood levels B, and K' and a are constants. This function can be linearized by taking the natural logarithm of both sides:

$$\ln P = \ln K' - aB$$
$$= K - aB$$
$$\text{where } K = \ln K'.$$

The constants K and a can be obtained by fitting the data points ln P and B to the linear function

$$\ln P = K - aB$$

This can be achieved by a least squares analysis, or by plotting ln P against B for a number of data points. Using the latter procedure, and forcing the line to pass through the $B = 20$, $\ln P = 3.0$ point (corresponding to $P = 20$), then a good fit occurs when

$$a = 0.03 \text{ and } K = 3.6, \text{ so } K' = 36.6$$

In other words, the appropriate exponential fit is

$$P = 36.6\, e^{-0.03B} \text{ for } B \geq 20.$$

The total percentage of children having $B > 100$ ppb can be determined by integrating this function from $B = 100$ to $B = \infty$.

$$\text{Total percentage} = \int_{100}^{\infty} P\,dB = 36.6 \int_{100}^{\infty} e^{-0.03B}\,dB$$

$$\text{Now} \int e^{ax}\,dx = e^{ax}/a, \text{ so}$$

$$\int_{100}^{\infty} e^{-0.03B}\,dB = -\frac{1}{0.03}\left[e^{-0.03B}\right]_{100}^{\infty}$$

$$= 33.3\,[0 - 0.0498]$$
$$= 1.658$$

Multiplying by 36.6, we find that the total percentage of children with levels > 100 equals 61%.

(b) Since the curve does not continue to rise, it implies that there are virtually no children with zero lead, so there are no environments that are completely lead-free. Instead, there is enough background lead to give almost every child a 20 ppb level.

4. There is no unique answer to a problem of this type; the objective is to estimate the order of magnitude of the requested quantity by finding first what pieces of information are required and then, in a commonsense approach, estimating values for the quantities whose values are not specified.

For convenience, consider a one-kilometer (0.6 mile) length of freeway; thus, the total area that we are considering is $1000\ m \times 2 \times 1000\ m = 2 \times 10^6\ m^2$. Consider first a single car that travels along this distance. If we estimate that at higher speeds, it can travel about 7 kilometers on one liter of gasoline (corresponding to about 20 miles per gallon), then it uses 1/7 liter of gasoline to travel the one kilometer, during which $0.2 \times 1/7$ grams of lead are emitted. Because this lead becomes spread over $2 \times 10^6\ m^2$, each car contributes $(0.2 / 7)\ g / 2 \times 10^6\ m^2$, i.e., 1.4×10^{-8} grams of lead per square meter.

If we assume that cars in each lane of the freeway on average pass by about every 10 seconds, there are six per lane per minute, or 360 per hour, or 8640 per day, or about 3×10^6 per year in each lane. Given that there are six lanes, the total cars per year is about 2×10^7. Thus, the amount of lead deposited per year on each square meter is:

$$1.4 \times 10^{-8}\ \text{grams / car / m}^2 \times 2 \times 10^7\ \text{cars / year} = 0.3\ \text{g / m}^2\text{ / year.}$$

Answers up to an order of magnitude greater than this volume (a *very* busy freeway), or an order of magnitude smaller, are reasonable.

5. (a) This problem can be solved by calculating the pH values at which each of the forms of the acids has a concentration equal to that of its conjugate base. For example, consider H_3AsO_4:

$$H_3AsO_4 \rightleftharpoons H_2AsO_4^- + H^+ \ (K_a = 6.3 \times 10^{-3})$$

$$\text{Thus,}\ \frac{[H_2AsO_4^-][H^+]}{[H_3AsO_4]} = 6.3 \times 10^{-3}$$

and so $[H^+] = 6.3 \times 10^{-3}\ [H_3AsO_4] / [H_2AsO_4^-]$

When $[H_3AsO_4] = [H_2AsO_4^-]$, then

$[H^+] = 6.3 \times 10^{-3}$ and $pH = 2.20$

At pH < 2.2, $[H_3AsO_4] > [H_2AsO_4^-]$, whereas the opposite is true for pH > 2.20.

Similarly, from the dissociation constant for $H_2AsO_4^-$, we find

$$[H_2AsO_4^-] = [HAsO_4^{2-}] \text{ at pH } = 6.89$$

and so $[H_2AsO_4^-] > [HAsO_4^{2-}]$ for pH < 6.89, etc.

Finally, from the dissociation constant for $HAsO_4^{2-}$, we find

$$[HAsO_4^{2-}] = [AsO_4^{3-}] \text{ at pH } = 11.49$$

Thus,

H_3AsO_4	predominates at	pH < 2.20
$H_2AsO_4^-$	predominates from	pH $= 2.20$ to 6.89
$HAsO_4^{2-}$	predominates from	pH $= 6.89$ to 11.49
AsO_4^{3-}	predominates at	pH > 11.49

Therefore, at pH $= 4$ and 6, $H_2AsO_4^-$ predominates, whereas at pH $= 8$ and 10, $HAsO_4^{2-}$ predominates.

(b) The equilibrium will be

$$H_3PO_3 \rightleftharpoons H^+ + H_2PO_3^-$$

$$\text{so } K_a = [H^+][H_2PO_3^-]/[H_3PO_3]$$

Solving for the required ratio, we obtain

$$[H_3PO_3] / [H_2PO_3^-] = [H^+]/K_a = [H^+]/6 \times 10^{-10}$$

When pH = 8, $[H^+] = 1 \times 10^{-8}$, so the ratio is $1 \times 10^{-8} / 6 \times 10^{-10} = 17$

When pH = 10, $[H^+] = 1 \times 10^{-10}$, so the ratio is $1 \times 10^{-10} / 6 \times 10^{-10} = 0.17$

6. If the metal's symbol in general is M, then its 2+ ion is M^{2+}. In each reaction, the charge on the metal species decreases by one because the charge on chloride is −1. Thus the general formulas are:

MCl^+, MCl_2 (no charge), MCl_3^-, and MCl_4^{2-}.

The formation reactions each involve chloride on the left side:

e.g., $MCl_2 + Cl^- \rightleftharpoons MCl_3^-$

Since seawater has a much higher concentration of chloride ion than freshwater, the equilibrium will be pushed much more to the right side in seawater. Thus complexes with 3 or 4 chlorines, which form only after 3 or 4 such reactions, would be found much more in seawater than in freshwater.

7. Mercury, lead, and cadmium are normally tied up in soil and rocks as the insoluble sulfide salts. However, acid rain provides the H^+ ions that by reaction to produce HS^- can release these metals into aqueous solution. Once liberated in this way, the metals can enter the food chain and water supply, and affect human health.

CHAPTER 13

Pesticides

Problem 13-1

Since free rotation occurs about the C—C bonds to the rings, the two ortho positions on a given ring are equivalent, and similar for the two meta positions. Thus the possible combinations are ortho, meta, and para on one ring, and the same for the other, yielding the combinations

ortho-ortho′	*o,o*′
ortho-meta and meta-ortho′	*o,m*′
ortho-para and para-ortho′	*o,p*′
meta-meta′	*m,m*′
meta-para and para-meta′	*m,p*′
(and para-para′	*p,p*′)

Problem 13-2

Since $\log K_{ow} = 5.7$

then $K_{ow} = 5.0 \times 10^5$

To a first approximation:

$$\frac{[S]_{fish\ fat}}{[S]_{water}} = K_{ow}$$

So,
$$\begin{aligned}[S]_{fish\ fat} &= K_{ow}[S]_{water} \\ &= 5.0 \times 10^5 \times 0.000010\ \text{ppm} \\ &= 5.0\ \text{ppm}\end{aligned}$$

Problem 13-3

Since DDD and DDT differ only in that the latter's —CCl_3 group is replaced in the former by —$CHCl_2$, the structure of DDD must be:

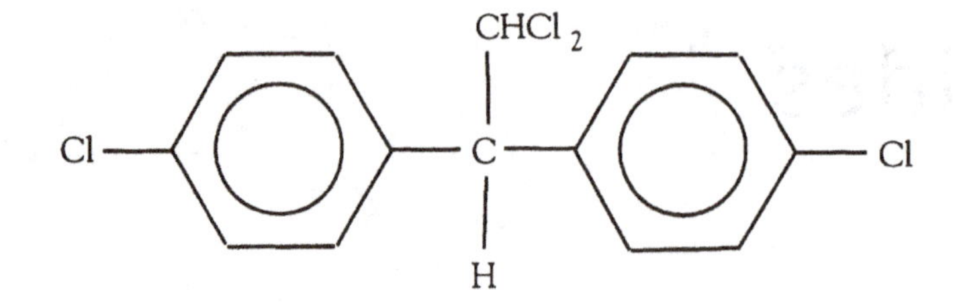

Problem 13-4

The insecticidal action depends on the size and shape; analogs must have the same size and shape as DDT to act in the same way. Thus:

(a) the substituted molecule containing the CH_3 groups instead of Cls will act as DDT does, because the Cl is substituted by a group of the same size.

(b) the substituted molecule containing H atoms will not act as DDT does, because the Cl is substituted here by a much smaller atom.

Problem 13-5

Doses on a mass of toxin per mass body weight basis are approximately transferrable between species, so first estimate the ratio of your mass to that of a mouse. Assuming that this is about 2000:1 (your mass being about 60 kilograms and that of the mouse being about 30 grams), then the mass that would kill you is 2000 × 0.2 micrograms = 400 micrograms. If you drink about 2 liters of water per day, then in 1 week you drink about 14 L. Thus the concentration in water would have to be 400 micrograms/14 L = 30 micrograms per liter, or 30 ppb.

Problem 13-6

Since the NOEL is 0.010 mg/kg/day and the usual safety factor is 100, the ADI or RfD is 0.010/100 = 0.00010 mg/kg/day. For a 55 kg woman, the maximum that should be ingested is 55 kg × 0.00010 mg/kg/day = 0.0055 mg/day.

Box 13-2, Problem 1

First we calculate the value of the fugacity, f, from the equation

$$f = n_{total} / \Sigma Z_x V_x$$

Since $n_{total} = 1$, and given the values supplied for the Z parameters and the model world volume parameters, we obtain in this case

$$\begin{aligned} f &= 1.0 / (4 \times 10^{-4} \times 10^{10} + 9.5 \times 10^{-5} \times 7 \times 10^{6} + 2.3 \times 2 \times 10^{4}) \\ &= 1.0 / (4 \times 10^{6} + 665 + 4.6 \times 10^{4}) \\ &= 2.47 \times 10^{-7} \text{ atm.} \end{aligned}$$

The concentrations of each chemical and the amounts of it can now be computed for each phase:

$$\begin{aligned} C_x = f Z_x \text{ so } & [\text{HCB in air}] = 2.47 \times 10^{-7} \times 4 \times 10^{-4} = 9.9 \times 10^{-11} \text{ mol / m}^3 \\ & [\text{HCB in water}] = 2.47 \times 10^{-7} \times 9.5 \times 10^{-5} = 2.3 \times 10^{-11} \text{ mol / m}^3 \\ & [\text{HCB in sediment}] = 2.47 \times 10^{-7} \times 2.3 = 5.7 \times 10^{-7} \text{ mol / m}^3 \end{aligned}$$

Box 13-2, Problem 2

First we calculate the value of the fugacity, f, from the equation

$$f = n_{total} / \Sigma Z_x V_x$$

Since $n_{total} = 1$, and given the values supplied for the Z parameters and the model world volume parameters, we obtain in this case

$$\begin{aligned} f &= 1.0 / (4 \times 10^{-4} \times 10^{10} + 2 \times 7 \times 10^{6} + 2 \times 10^{-5} \times 2 \times 10^{4}) \\ &= 1.0 / (4 \times 10^{6} + 1.4 \times 10^{7} + 0.4) \\ &= 5.56 \times 10^{-8} \text{ atm.} \end{aligned}$$

The concentrations of each chemical and the amounts of it can now be computed for each phase:

$$\begin{aligned} C_x = f Z_x \text{ so } & [\text{Dieldrin in air}] = 5.56 \times 10^{-8} \times 4 \times 10^{-4} = 2.2 \times 10^{-11} \text{ mol / m}^3 \\ & [\text{Dieldrin in water}] = 5.56 \times 10^{-8} \times 2 = 1.1 \times 10^{-7} \text{ mol / m}^3 \\ & [\text{Dieldrin in sediment}] = 5.56 \times 10^{-8} \times 2 \times 10^{-5} = 1.1 \times 10^{-12} \text{ mol / m}^3 \end{aligned}$$

The amounts in each phase are given by the f Z V values. Thus, the number of moles of Dieldrin

$$\begin{aligned} &\text{in air} = 2.2 \times 10^{-11} \times 1 \times 10^{10} = 0.22 \text{ mol} \\ &\text{in water} = 1.1 \times 10^{-7} \times 7 \times 10^{6} = 0.77 \text{ mol} \\ &\text{in sediments} = 1.1 \times 10^{-12} \times 2 \times 10^{4} = 2 \times 10^{-8} \text{ mol} \end{aligned}$$

Thus most of the dieldrin will be found in water, the rest in air.

Green Chemistry Problems

1. (a) 3. The design chemicals are less toxic than current alternatives or inherently safer with regard to accident potential.

   (b) 4. Preserving efficacy of function while reducing toxicity.

2. They are specific to certain target organisms and thus do not harm beneficial insects.

3. (a) Under this program, which was started in 1993, a pesticide must meet one or more of the following requirements. It must

   (1) reduce pesticide risks to human health;
   (2) reduce pesticide risks to non-target organisms;
   (3) reduce the potential for contamination of valued environmental resources;
   (4) broaden adoption of IPM (integrated pest management) or make it more effective.

   (b) 1 and 2.

4. They target specific insects by interrupting a physiological process that is unique to these insects. In the case of Confirm, Mach2, and Intrepid, they disrupt the process of molting.

5. (a) 3. The design chemicals that are less toxic than current alternatives or inherently safer with regard to accident potential.

   (c) 1. Prevention of waste.

   4. Preserving efficacy of function while reducing toxicity.

6. Hexaflumuron is generally less toxic to mammals than traditional pesticides, and the Sentricon system allows for the use of from 100 to 1000 times less pesticide than conventional treatment methods.

7. 1, 2, and 3.

8. (a) 3. The design of chemicals that are less toxic than current alternatives or inherently safer with regard to accident potential.

   (b) 1. Prevention of waste.

   3. Wherever practicable, synthetic methodologies should be designed to use and generate substances that possess little or no toxicity to human health and the environment.

   4. Chemical products should be designed to preserve efficacy of function while reducing toxicity.

   7. A raw material feedstock should be renewable rather than depleting whenever technically and economically practical.

9. The reduction in toxicity to non-target species, prolonged efficacy, renewable feedstocks, and reduction in energy.

10. The ring containing the ester functional group.

Additional Problems

1. The concentration of DDE in the breast milk in 1972 was *ca.* 700 g per 10^9 g lipid (fat).

   ∴ Mass of DDE in 250 mL of breast milk = 250 mL × 4.2 g fat/100mL × 700 g DDE/10^9 g fat = 7.4×10^{-6} g DDE = 7.4 μg DDE

2. BCF = K_{ow} × % body fat / 100

   From Table 13-3, log K_{ow} = 3.8 for parathion, ∴ $K_{ow} = 10^{3.8} = 6300$

   ∴ BCF = 6300 × 0.05 = 315

   $$\text{BCF} = \frac{\text{concentration of parathion in fish}}{\text{concentration of parathion in lake}}$$

   ∴ Concentration of parathion in lake = concentration of parathion in fish/BCF
   = 22 ppb / 315
   = 0.070 ppb

3. If the NOEL is 0.004 mg/kg, then using the safety factor of 100, the ADI/RfD would be 0.00004 mg/kg. Assuming a body mass of 60 kg, the maximum daily exposure should not exceed 60 kg body weight × 0.00004 mg chemical/1 kg body weight = 2.4×10^{-3} mg of the chemical. Since in fish the average concentration is 0.2 ppm, there exists 0.2 grams of the chemical per million grams of fish. Thus the mass of fish containing 2.4×10^{-3} mg is

   $$2.4 \times 10^{-3} \text{ milligrams chemical} \times \frac{1 \text{ g chemical}}{1000 \text{ mg chemical}} \times \frac{10^6 \text{ g fish}}{0.2 \text{ grams chemical}}$$

   = 12 g fish

4. (a) Using d = 0, 1, ...5 and d = 0.5 for the linear d scale, the form of the curve is the same as that in Figure 13-4a. Using d = 0.01, 0.02, ...0.10, 0.20, 0.5, 1, 2, ...5, and taking ln d to obtain values of –4.60, –3.91, ...and plotting R versus ln d, we obtain a curve similar in form to Figure 13-4b.

   (b) From the equation

   $$R = 1 - e^{-d}$$

   we can substitute R = 0.5 for the LD_{50} dose, and solve for d:

   $$0.5 = 1 - e^{-d}$$
   $$\therefore e^{-d} = 0.5$$
   $$-d = \ln(0.5) = -0.693$$
   so d = 0.69 from the equation.

   From the graphs, $d \sim 0.7$ from both curves.

(c) According to the equation, $R = 0$ corresponds to $e^{-d} = 1$, so $d = 0$ and no threshold is predicted. It is difficult to tell visually from the graph for the log function whether a threshold exists or not.

5. Mass azinphos-methyl in lake = 200 g × 0.35 = 70 g

Volume of water in the lake = 30,000 m^2 × 0.5 m = 15,000 m^3

Mass of water in the lake = 15,000 m^3 × 10^6 cm^3 / m^3 × 1.0 g / cm^3 = 1.5 × 10^{10} g

∴ Concentration of azinphos-methyl in lake = 70 g / 1.5 × 10^{10} g water = 4.7 × 10^{-9} g / g H_2O = 4.7 ppb

This concentration is significantly above the LC_{50}, so we would expect that a major fish kill would have occurred.

CHAPTER 14

Dioxins, Furans, and PCBs

Problem 14-1

Since the 1, 4, 6 and 9 positions of the dibenzo-*p*-dioxin ring are all equivalent, as are the 2, 3 ,7, and 8 positions, the 1, 3 and 2, 4 and 6, 8 and 7, 9 - dichlorodibenzodioxins are all the same molecule (which is properly numbered as 1, 3 -).

Both 1, 2 - and 1, 8 - dichloro derivatives are unique because, although in the unsubstituted dibenzodioxin its 2 and 8 positions are equivalent, this is no longer true after the 1 - position is substituted (consult the numbered dioxin structure in the text to convince yourself of this).

To generate in turn all the dichlorodioxins, first place one Cl at C-1 and successively place the second chlorine at the other positions. Compare each new structure to previous ones for uniqueness by mentally rotating each by 180° about the O—O line and also by flipping it over.

The unique dichlorodibenzodioxins are those numbered:

1, 2	
1, 3	2, 3
1, 4	(not 2, 4 as this is $\equiv$ 1, 3)
1, 6	(not 2, 6 as this is $\equiv$ 1, 7)
1, 7	2, 7
1, 8	2, 8
1, 9	(not 2, 9 as this is $\equiv$ 1, 8)

Problem 14-2

In both OCDD and pentachorophenol, all hydrogens have been substituted and thus there is no question of the specific positions they occupy or of isomers that need to be distinguished. In contrast, in 2, 3, 7, 8 -TCDD, only 4 of the 8 positions are occupied by chlorine and thus isomers are possible and must be distinguished by specifying the locations of the Cl atoms.

Problem 14-3

The remaining H atom on the ring can occupy the position ortho, meta, or para to the OH group. It is in the meta position in the 2, 3, 4, 6 isomer. Hence the other two isomers have the structures below:

(a)

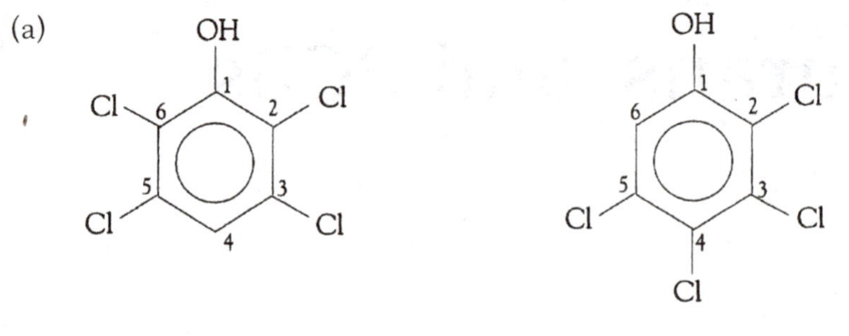

2, 3, 5, 6 - tetrachlorophenol 2, 3, 4, 5 - tetrachlorophenol

(b) Align the phenols so the —OH groups are in the region between the rings, and each opposite a chlorine ready to join them together:

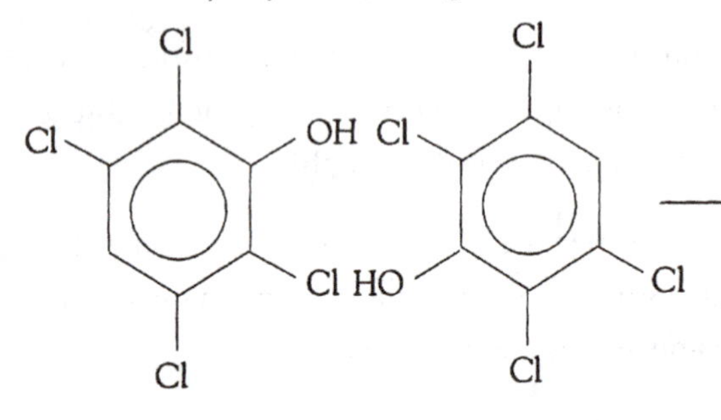

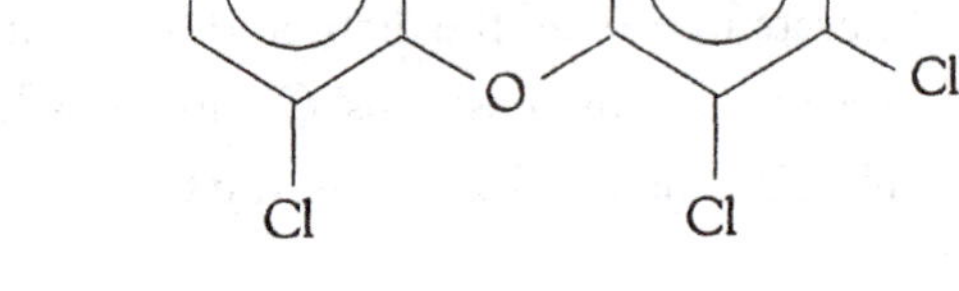

(Both isomers are 2, 3, 5, 6 - tetrachlorophenol.) 1, 2, 4, 6, 7, 9 - hexachlorodibenzodioxin

This is the only relative orientation possible.

Next, consider the interaction of two molecules of the 2, 3, 4, 5 isomer of tetrachlorophenol:

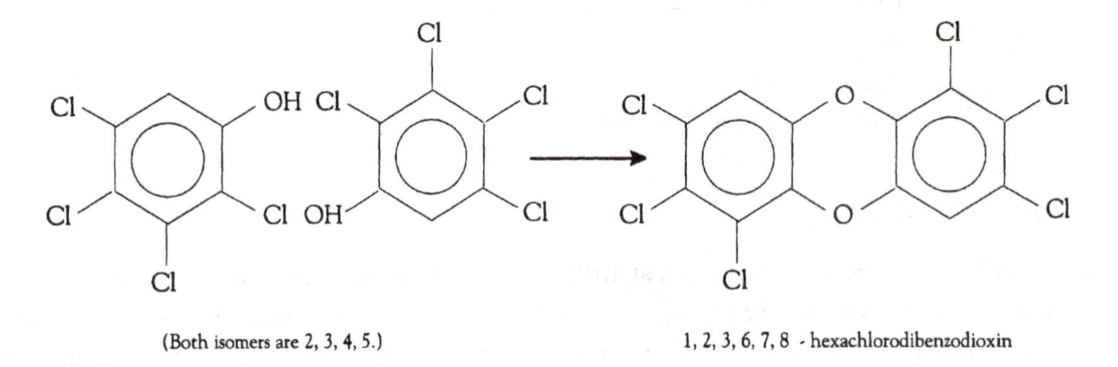

(Both isomers are 2, 3, 4, 5.) 1, 2, 3, 6, 7, 8 - hexachlorodibenzodioxin

Again, this is the only relative orientation possible because there must be Cl's pointing at the —OH's in the opposite ring.

Problem 14-4

"Dioxins" are produced by the coupling of two chlorophenols, as illustrated in Problem 14-3b.

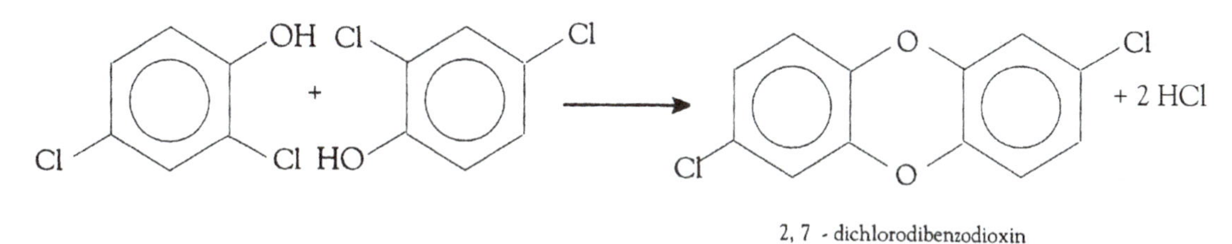

2, 7 - dichlorodibenzodioxin

Problem 14-5

From its structure, we can determine the molecular formula of 2, 3, 7, 8 - TCDD to be $C_{12}H_4O_2Cl_4$, for which we obtain its molar mass of 321.96 g / mol. Thus, the number of moles of TCDD in a sample of 10^{-16} grams of it is

$$10^{-16} \text{ grams} \times 1 \text{ mole} / 321.96 \text{ grams} = 3.11 \times 10^{-19} \text{ moles}$$

Thus the number of molecules in the sample is

$$3.11 \times 10^{-19} \text{ moles} \times 6.02 \times 10^{23} \text{ molecules / mole} = 1.87 \times 10^{5} \text{ molecules}$$

Problem 14-6

Consider one mole of $C_{12}H_{10-n}Cl_n$; the mass of chlorine in it is $35.45n$ because the atomic mass of chlorine is 35.45, whereas its total molar mass is:

$$12.01 \times 12 + 1.01\,(10 - n) + 35.45n$$

Thus, the fraction of the PCB that is chlorine is

$$\frac{35.45n}{12.01 \times 12 + 1.01\,(10 - n) + 35.45n} = 0.60$$

Rearranging this equation using standard methods of algebra,

$$35.45n = 86.47 + 6.06 - 0.61n + 21.27n$$

$$14.79n = 92.53$$

$$n = 6.26$$

Problem 14-7

The unique dichlorobiphenyls, assuming no rotation about C—C, would be numbered as follows:

Both Cl's on the same ring:

2, 3
2, 4
2, 5
2, 6
3, 4
3, 5
(but not 3, 6 as it is identical to 2, 5)

One Cl on each ring

2, 2'	and 2, 6' only if there is no free rotation
2, 3'	and 2, 5' only if there is no free rotation
2, 4'	
3, 3'	and 3, 5' only if there is no free rotation
3, 4'	
4, 4'	

Problem 14-8

As stated in the Hint, the fraction f of the initial amount left after time t has elapsed is given by

$$f = e^{-kt}$$

where here the rate constant k equals 0.078 yr^{-1}. We can take 1994 to be equivalent to $t = 0$, so since 2010 is 16 years beyond this date, $t = 16$ yr. Thus

$$f = e^{-0.078 \times 16} = e^{-1.248} = 0.287$$

Thus since the initial concentration was 0.047 ppt, the projected concentration in 2010 is $0.287 \times 0.047 = 0.013$ ppt.

To deduce the year when the concentration is 0.010, first determine the corresponding f value (relative to 1994):

$$f = 0.010 / 0.047 = 0.213$$

Thus $0.213 = e^{-0.078t}$

We can linearize this equation by taking the natural logarithm of both sides:

$$\ln(0.213) = -0.078\,t$$

$$-1.548 = -0.078\,t$$

Solving for t, we obtain $t = 19.8$

Problem 14-9

The structure and numbering for the rings are:

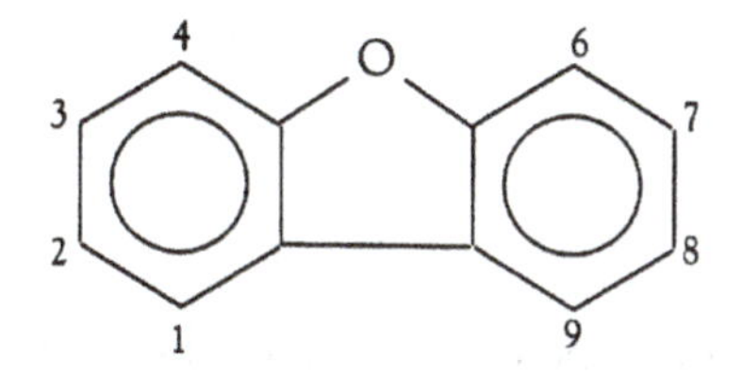

Thus, for example, 1, 2-dichlorodibenzofuran has the structure:

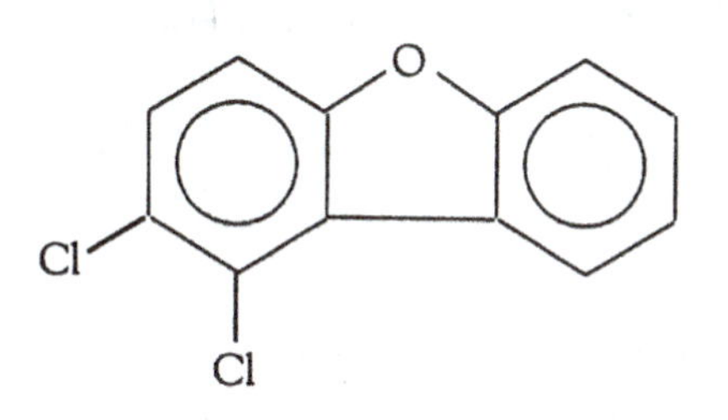

As in Problem 14-1, to generate all the dichlorofurans, start with one chlorine at C-1 and in turn place the second chlorine at each other position. Then place the chlorine at C-2 and again generate all possibilities, etc. In each case, check each new structure for possible equivalence to ones generated previously.

The unique dichlorofurans are listed below:

Both Cl atoms on the same ring:

1,2
1,3
1,4
2,3
2,4 (not identical to 1,3 since furans do not have "up-down" symmetry)
3,4 (not identical to 1,2 since furans do not have "up-down" symmetry)

One Cl on each ring:

1,6
1,7
1,8
1,9
2,6 (not identical to 1,7 since furans do not have "up-down" symmetry)
2,7
2,8 (but not 2,9 since it is identical to 1,8)
3,6
3,7 (but not 3,8 or 3,9 since they are identical to 2,7 and 1,7)
4,6 (but not 4,7 or 4,8 or 4,9)

Problem 14-10

First, it is useful to list the number of alpha and beta chlorine substituents because toxicity is dependent upon them:

Dioxin	*# alpha Cl (1, 4, 6, 9)*	*# beta Cl (2, 3, 7, 8)*
2, 3, 7 -	0	3
1, 2, 3 -	1	2
1, 2, 3, 7, 8 -	1	4

The dominant factor in determining dioxin toxicity is the number of beta chlorines minus the number of alpha chlorines, with the beta effect dominating somewhat the alpha. The number of betas minus the number of alphas is 3 for both 2,3,7- and 1, 2, 3, 7, 8-congeners, but we predict the latter to be more toxic since the beta effect is stronger. Both should be more toxic than the 1,2,3-congener, since for it the number of beta minus alpha is only 1.

Toxicity order: 1, 2, 3, 7, 8 > 2, 3, 7 > 1, 2, 3

Problem 14-11

The contribution of each compound equals its actual mass times the equivalency factor in Table 14-1; thus:

$$\text{Equivalent mass} = 24\text{ pg} \times 0.1 + 52\text{ pg} \times 0.5 + 200\text{ pg} \times 0.001$$
$$= 28.6\text{ pg of 2, 3, 7, 8 - TCDD}$$

Problem 14-12

Recall that 1 picogram is 10^{-12} grams, so that food contains about 0.4×10^{-12} grams of TCDD per gram. Assuming that your body mass is about 60 kg, a fatal dose of 2, 3, 7, 8 - TCDD is about $60 \text{ kg} \times 0.001 \text{ mg/kg} = 0.06 \text{ mg} = 0.06 \times 10^{-3}$ grams. Thus the mass of food required is

$$\frac{0.06 \times 10^{-3} \text{g TCDD}}{0.4 \times 10^{-12} \text{g TCDD/1 g food}} = 1.5 \times 10^{8} \text{g food}$$

The mass of animal-based food required is about 1.5×10^{5} kilograms.

Box 14-1, Problem 1

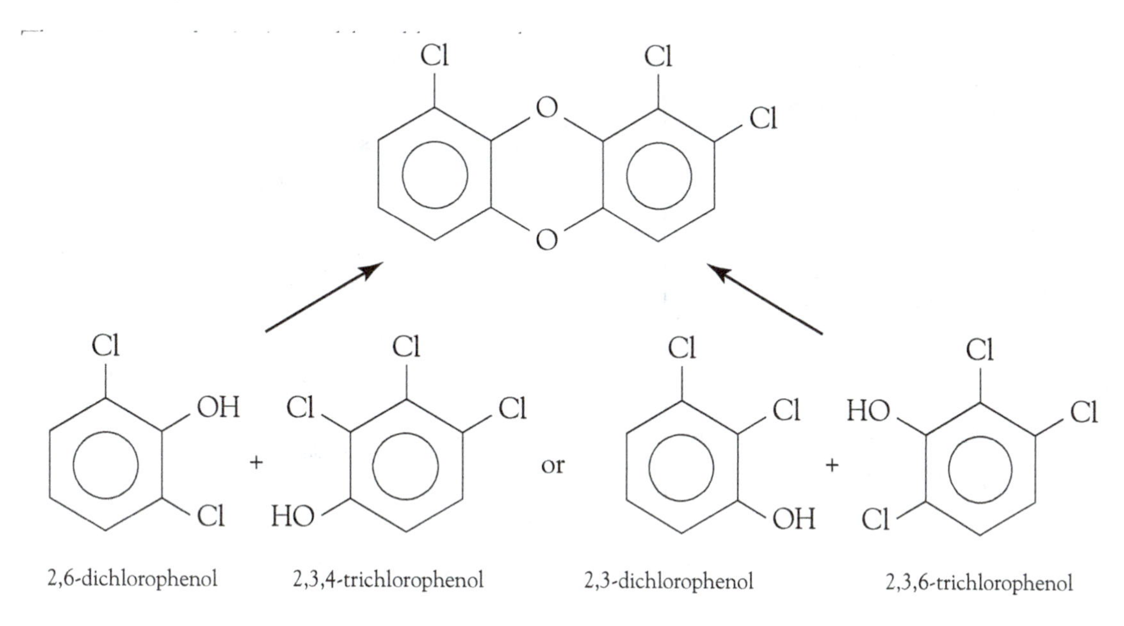

As shown above, it can be formed in two ways depending upon which oxygen is initially associated with which chlorophenol molecule.

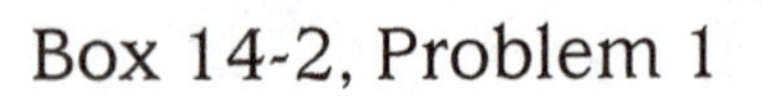

Box 14-2, Problem 1

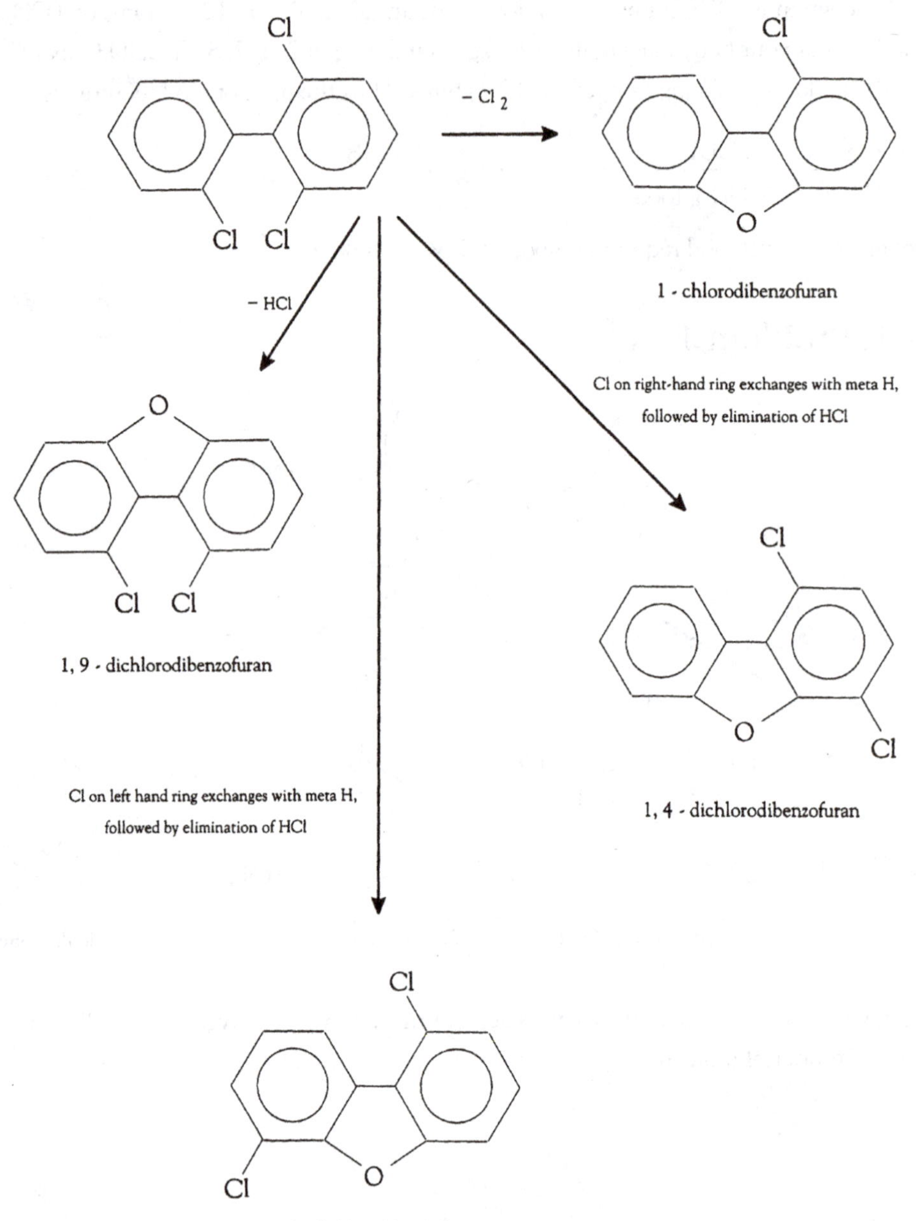

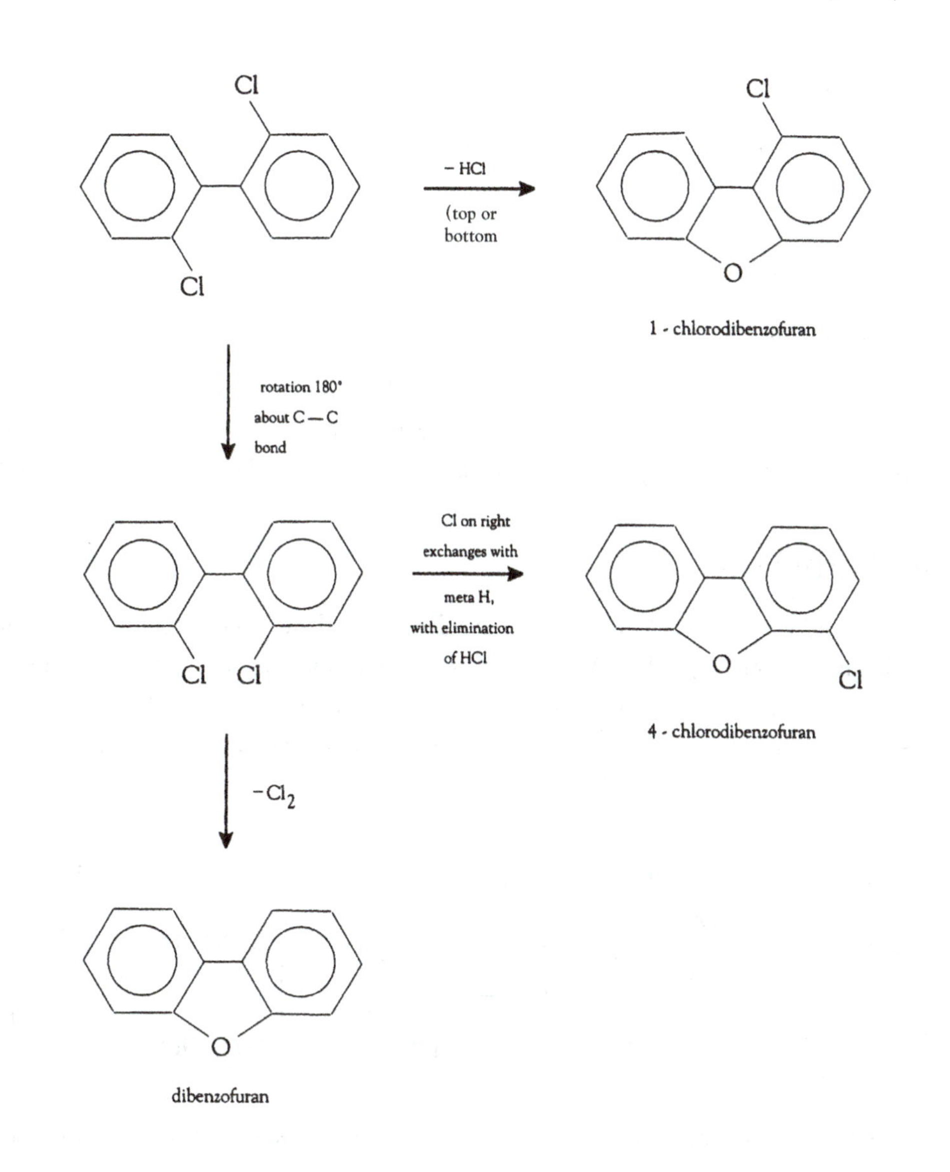

Box 14-2, Problem 2

No other furans would be produced from the particular trichlorobiphenyl illustrated, as it contains no ortho hydrogen atom pair. In the case of the dichlorobiphenyl, if rotation about the C—C bond first occurs, then a pair of ortho hydrogens would occur on the same side of the molecule, and could be eliminated as H_2:

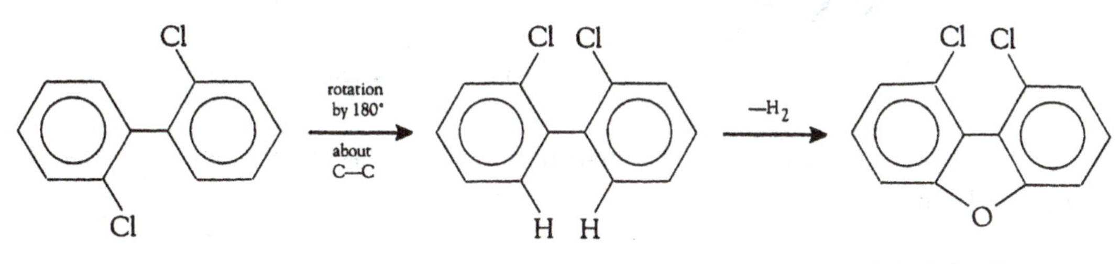

1, 9 - dichlorodibenzofuran

Green Chemistry Problems

1. (a) 2. Alternative reaction conditions for green chemistry.

   (b) 1. Prevention of waste.

   4. Preserving efficacy of function while reducing toxicity.

   6. Energy reduction.

2. Formation of chlorinated organic compounds such as dioxins is eliminated. There is also the possibility of using lower temperatures so less energy would be needed and better selectivity would be obtained, leading to less destruction of cellulose fibers and therefore less wood needed.

Additional Problems

1. Use a "retrosynthesis" approach to solve this problem—i.e., deduce a possible set of reactants that could give a specified product. When a dibenzodioxin ring is formed, an —OH and an adjacent —Cl react; the former displaces —Cl on the adjacent ring, and the latter combines with —O of the second ring. New bonds joining rings are indicated by arrows below:

(a)

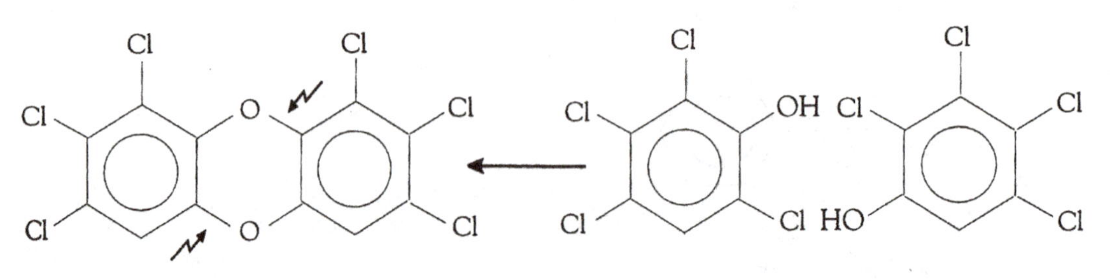

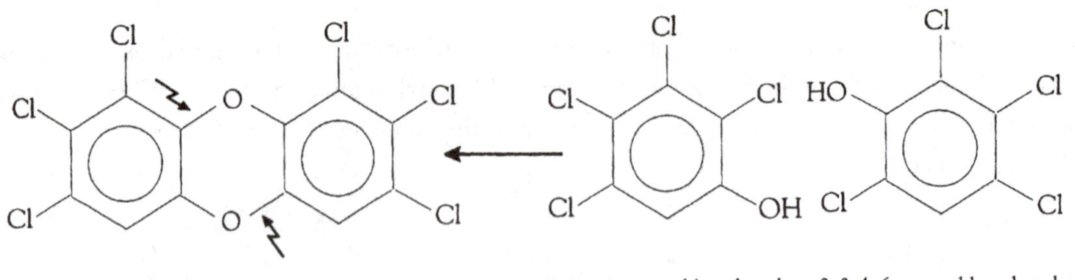

Thus, the combination of one 2, 3, 4, 5 - and one 2, 3, 4, 6 - tetrachlorophenol molecule leads to only one dioxin.

(b)

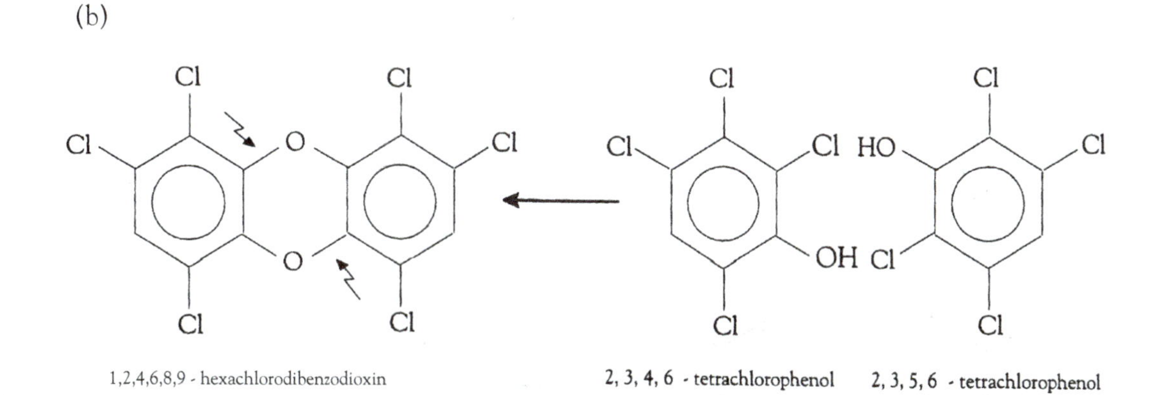

The alternative set of new bonds gives the same two tetrachlorophenols as shown above at right.

(c)

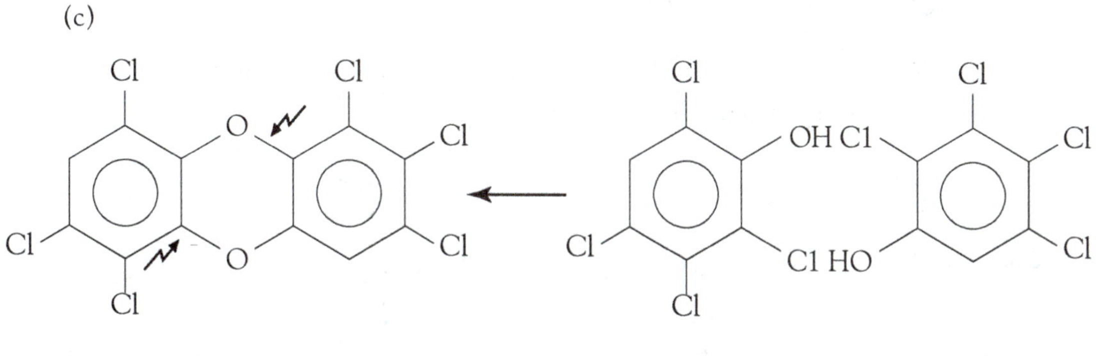

The alternative set of new bonds arises from coupling 2, 3, 4, 6 - and 2, 3, 5, 6 - tetrachlorophenol.

2. "Commercial" PCP contains not only pentachlorophenol but also 2, 3, 4, 6 - tetrachlorophenol (as stated in text). Thus, we must consider, besides the coupling of two PCP molecules, (i) of two tetrachlorophenol molecules, and (ii) a combination of one of each type of molecule:

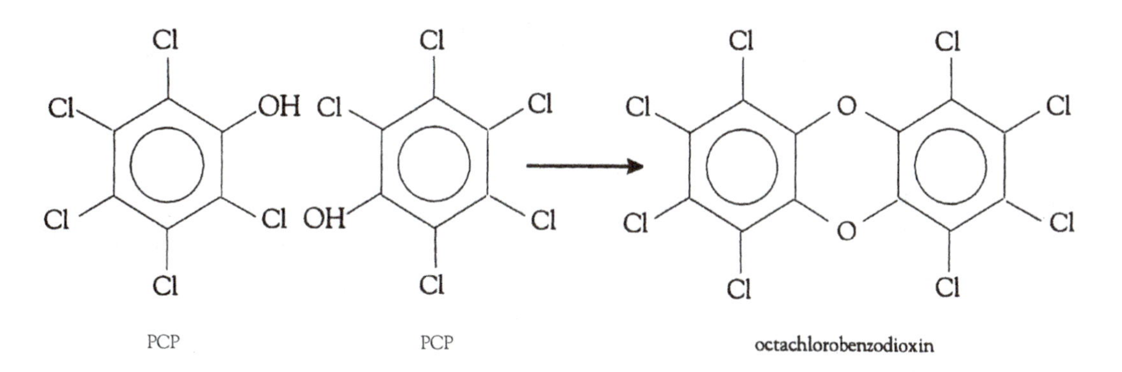

Two 2, 3, 4, 6 - tetrachlorophenols ⟶ 1, 2, 3, 6, 8, 9 - hexachlorodibenzodioxin
and 1, 2, 4, 6, 7, 9 -
and 1, 2, 3, 6, 7, 8 -

(ii) one of each:

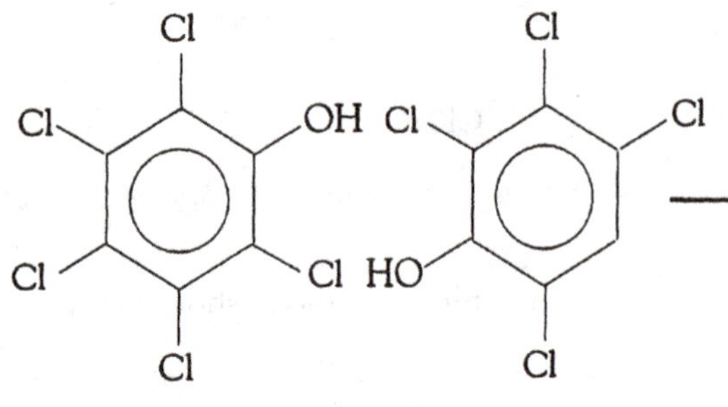

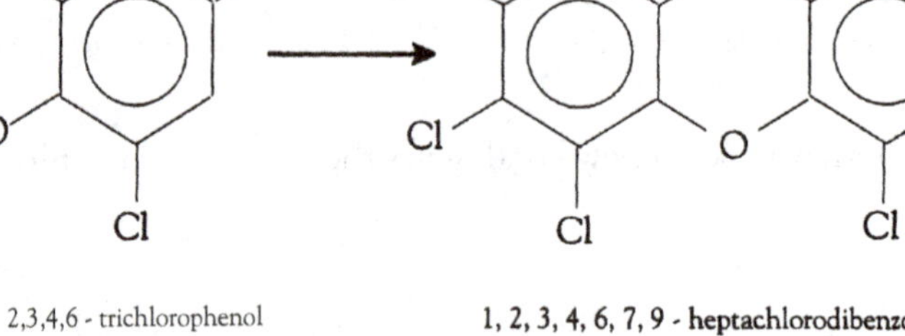

PCP   2,3,4,6 - trichlorophenol   1, 2, 3, 4, 6, 7, 9 - heptachlorodibenzodioxin

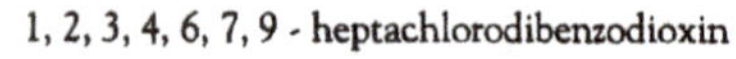

or via opposite alignment of second ring:

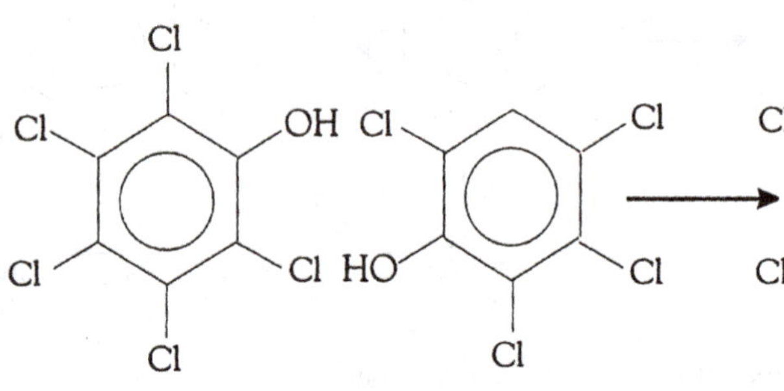

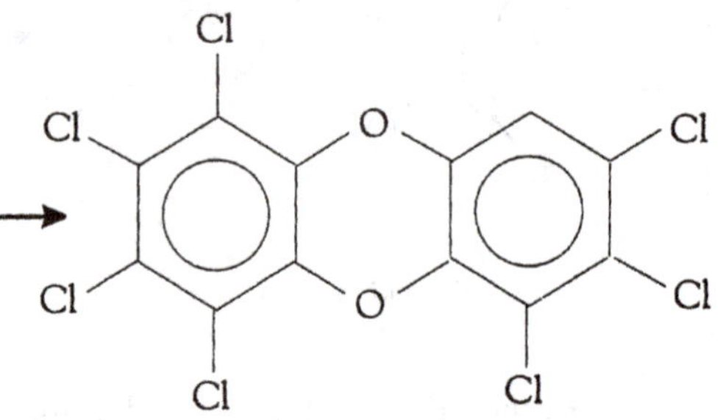

1, 2, 3, 4, 6, 7, 8 - heptachlorodibenzodioxin

3. Consider the possible coupling of pentachlorophenol with 2, 3, 5, 6 - tetrachlorophenol by loss of 2 HCl:

One orientation of the phenols is:

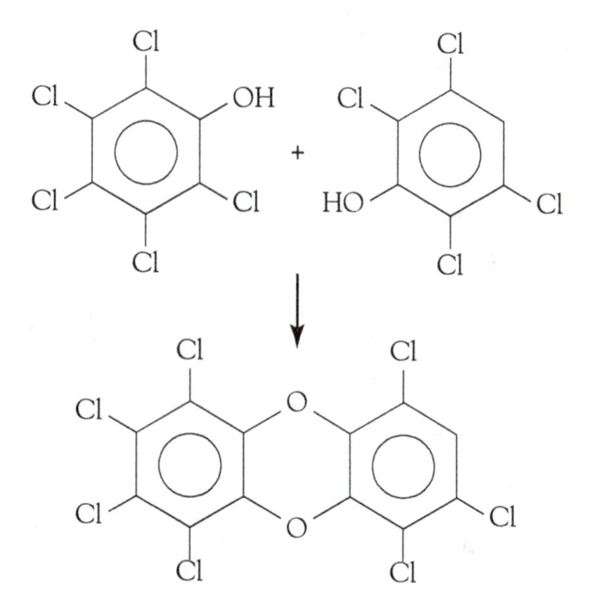

1, 2, 3, 4, 6, 7, 9 - heptachlorodibenzo-*p*-dioxin

There is no other unique orientation of the phenols. Coupling two pentachlorophenols would give an octachlorodibenzo-*p*-dioxin, and that of two tetrachlorophenols necessarily gives a hexachlorodibenzo-*p*-dioxin since 2 Cl's would be eliminated. Thus the 1, 2, 3, 4, 6, 7, 8 - heptachlorodibenzo-*p*-dioxin must have originated by dechlorination of OCDD.

The flaw is that dechlorination of highly chlorinated dioxins will such as OCDD yield more toxic, medium-chlorinated dioxins.

5. Using a "retro-synthesis" approach as in Additional Problem 1:

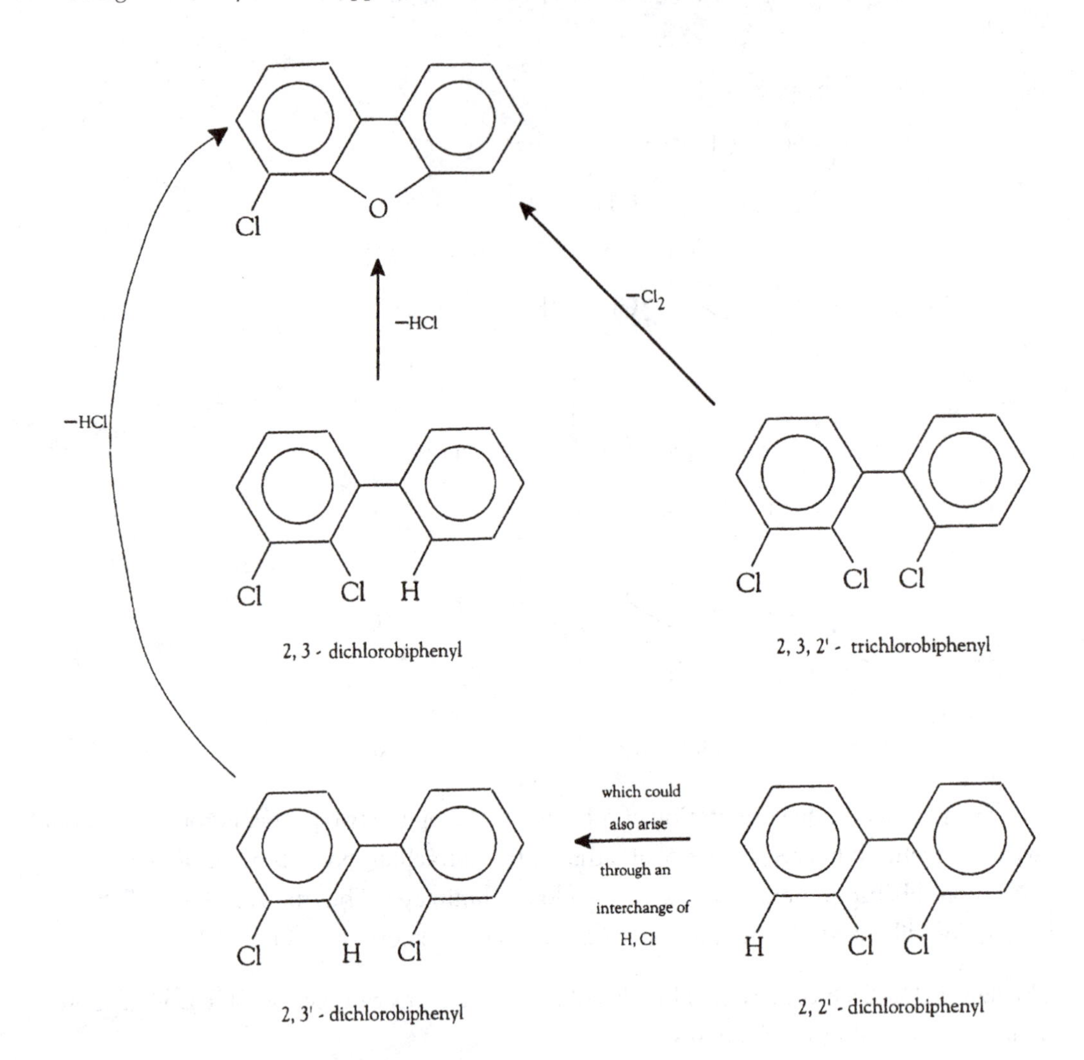

Thus, 2,2' and 2,3' and 2,3,2' congeners could produce that furan.

6. (a) Since the average North American has a TEQ level in his/her fat of about 40 ppt, and since body fat represents far more than 1% of body mass, the average concentration in the body is far greater than 0.4 ppt, the average level in food. Thus humans apparently do bio-magnify dioxins.

   (b) The levels for fish exceed those for domestic animals because their food chain is longer and so there are more stages at which the dioxin can be biomagnified.

   (c) The TEQ level for butter is higher because it contains a much higher proportion of fat than does milk, and similar for hot dogs compared to other meat.

(d) The vegan TEQ value is very low because non-meat, non-dairy products contain much lower levels of dioxins, etc. than do products containing animal fat. Given that the vegan TEQ is much less than that for animals, some biomagnification must occur for animals.

7. The key factor is the presence of ortho Cl atoms, and the resulting inability of the PCB to adopt a coplanar geometry of the two aromatic rings (thus giving a shape similar to 2, 3, 7, 8 - TCDD). First, draw the three PCB congeners:

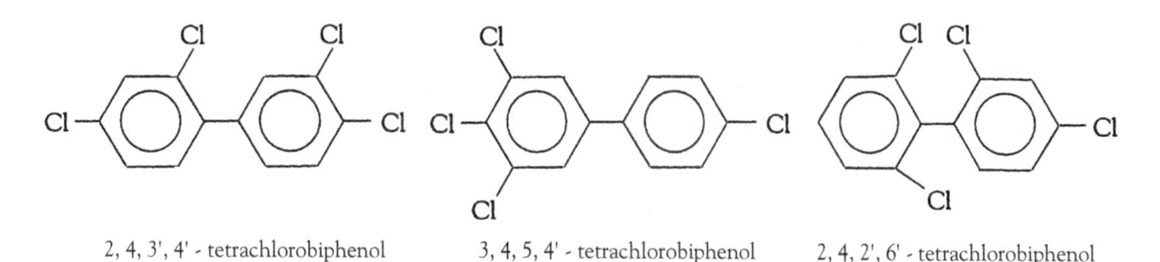

The most toxic is 3, 4, 5, 3', since it has no ortho Cl atoms and thus can easily adopt a coplanar geometry (with complete rotational freedom about the C-C single bond), and also because it has meta and para chlorines, which contribute to toxicity. The least toxic is 2, 4, 2', 6', since it has Cl atoms substituted in three of the four ortho positions, and thus is completely prevented from adopting a coplanar geometry. The 2, 4, 3' ,4' congener is intermediate; it has only one ortho Cl atom, and thus has hindered rotation, but still should be able to adopt a coplanar geometry.

8. *2, 6 - dichlorophenol:*

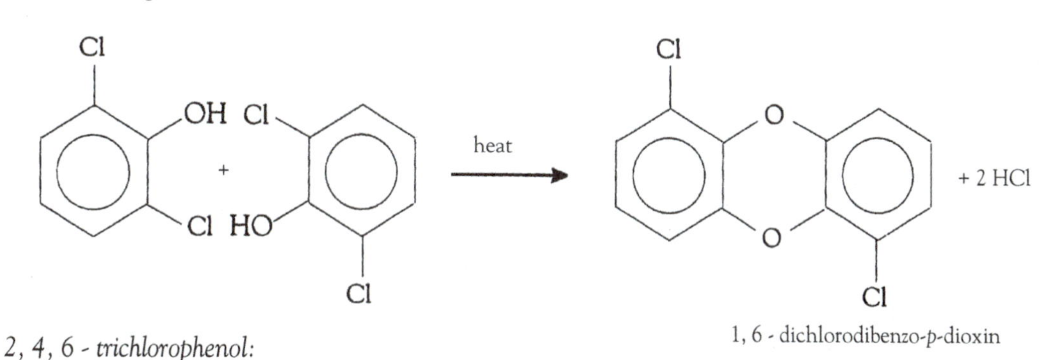

2, 4, 6 - trichlorophenol:

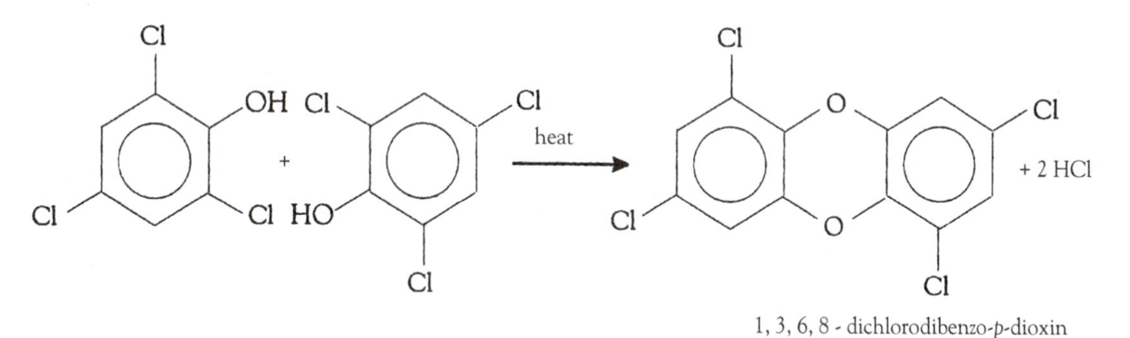

Due to the symmetry of these two chlorophenols, only one possible PCDD can be formed in each case. In terms of toxicity, dichlorodioxins are not usually considered to be highly toxic. This is especially true in this case of 1, 6 - dichlorobenzo-*p*-dioxin, which contains only α and no β chlorines. 1, 3, 6, 8 - terachlorodibenzo-*p*-dioxin would be more toxic, since it does have two β chlorines.

9. First we calculate the value of the fugacity, f, from the equation

$$f = n_{total} / \Sigma Z_x V_x$$

Since $n_{total} = 1$, and given the values supplied for the Z parameters and the model world volume parameters, we obtain in this case

$$\begin{aligned} f &= 1.0 / (4 \times 10^{-4} \times 10^{10} + 0.03 \times 7 \times 10^{6} + 10000 \times 2 \times 10^{4}) \\ &= 1.0 / (4 \times 10^{6} + 2.1 \times 10^{5} + 2.0 \times 10^{8}) \\ &= 4.90 \times 10^{-9} \text{ atm.} \end{aligned}$$

The concentrations of each chemical and the amounts of it can now be computed for each phase:

$C_x = f Z_x$ so [PCBs in air] $= 4.90 \times 10^{-9} \times 4 \times 10^{-4} = 2.0 \times 10^{-12}$ mol / m^3
[PCBs in water] $= 4.90 \times 10^{-9} \times 0.03 = 1.5 \times 10^{-10}$ mol / m^3
[PCBs in sediment] $= 4.90 \times 10^{-9} \times 10000 = 4.9 \times 10^{-5}$ mol / m^3

Notice the preferential concentration of PCBs in sediment.

CHAPTER **15**

Other Toxic Organic Compounds of Environmental Concern

Problem 15-1

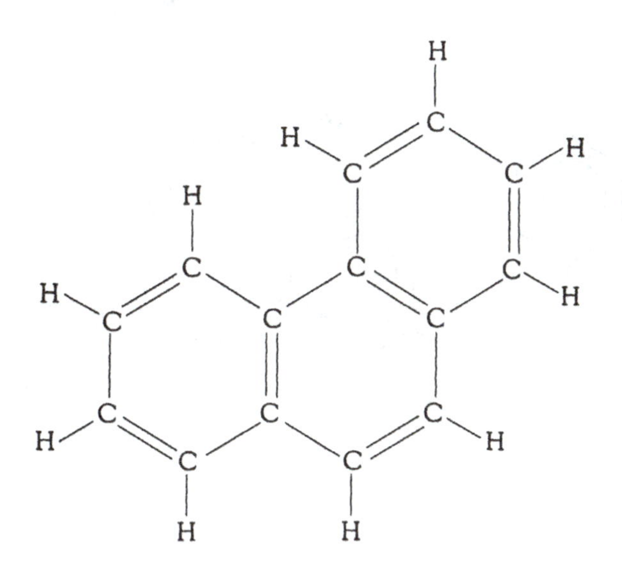

(Other contributing resonance structures also exist.)

Problem 15-2

The molecule at the left is just phenanthrene rotated around in its plane by 180°.

The molecule at the right is not even an isomer, because it has only 13 rather than 14 carbons, as in phenanthrene.

Problem 15-3

To deduce all unique isomers, consider the centers of the C_6 rings to be points and generate all possible unique arrangements.

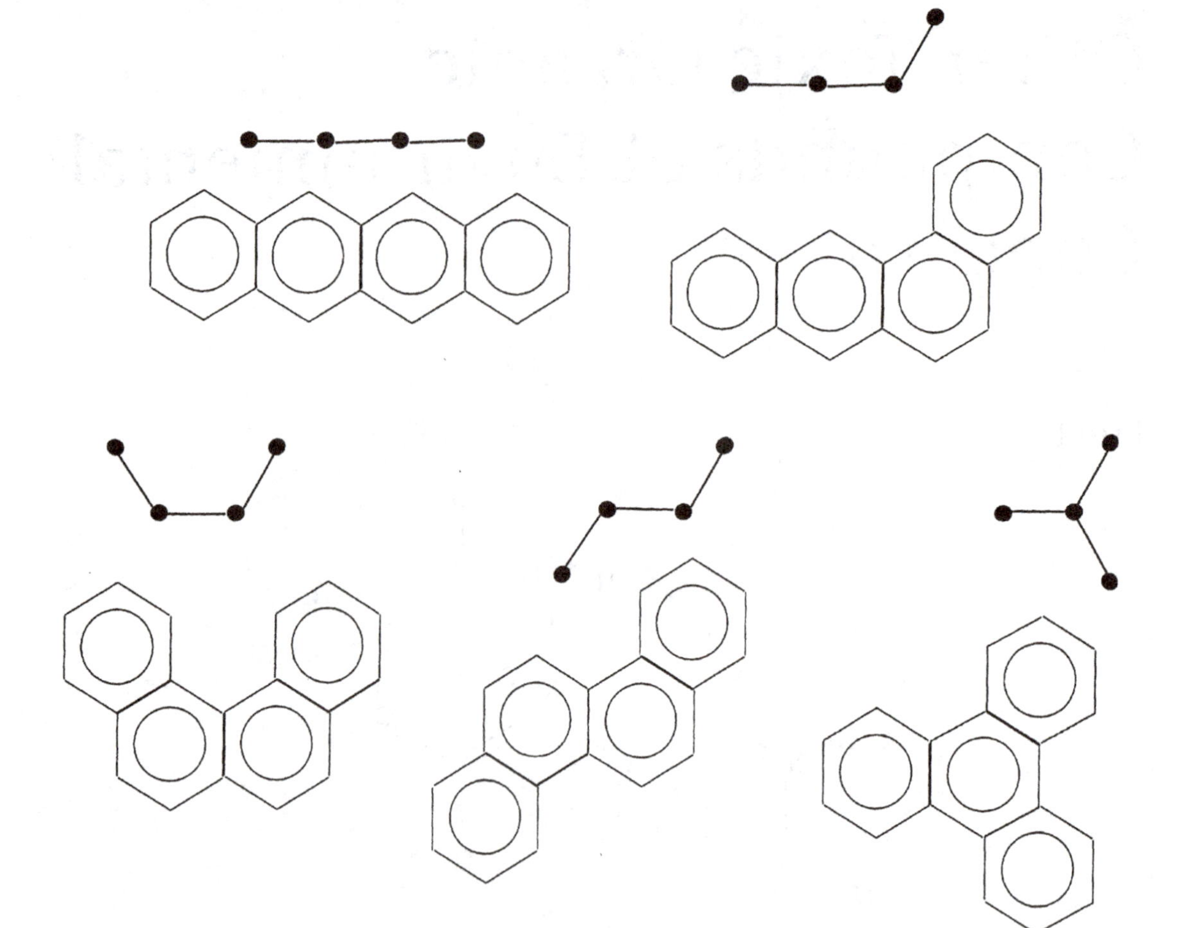

Problem 15-4

Naphthalene:	No, since it does not have a bay region.
Anthracene:	No, since it does not have a bay region.
Phenanthrene:	Yes, since it has at least one bay region.
Benzo[ghi]perylene:	Yes, since it has at least one bay region.

Problem 15-5

Given the data for DDT in the text, we see that DDE has a smaller vapor pressure (0.0032 vs. 0.005) and a lower condensation temperature (–2°C vs. 13°C).

Since mobility (once evaporated) depends upon condensation temperature, more of it will be deposited further toward the poles, i.e., at higher latitudes.

Problem 15-6

(a) Since the two rings in decabromo diphenylether are equivalent and free rotation is stated not to occur around the C-O bonds, for simplicity the structures can be drawn in the linear fashion shown below. (For simplicity, the bromine atoms are not shown.) The unique positions from which one Br could be removed are those ortho, meta, and para relative to the carbon attached to the oxygen, as indicated by arrows:

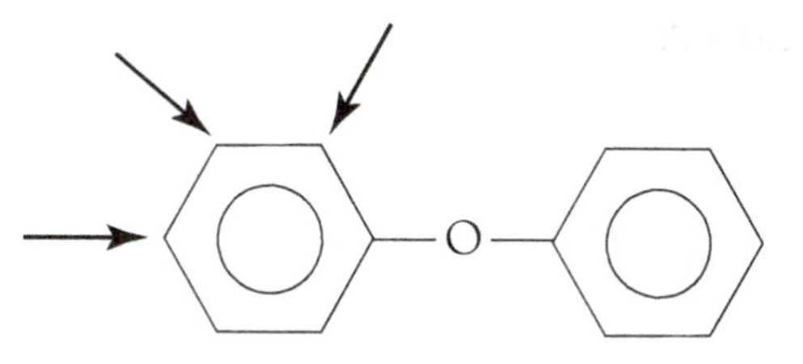

(b) Using a systematic procedure, consider the possible positions in each ring for the removal of the second bromine (shown as a dot) for each of the possibilities of one missing bromine (arrow) discovered in part (a). Note that when the first bromine is removed from an ortho or meta position in the ring at the left side, due to the lack of free rotation about the C-O bonds there exist two separate isomers in which the second bromine is missing from ortho and meta positions on the right-side ring:

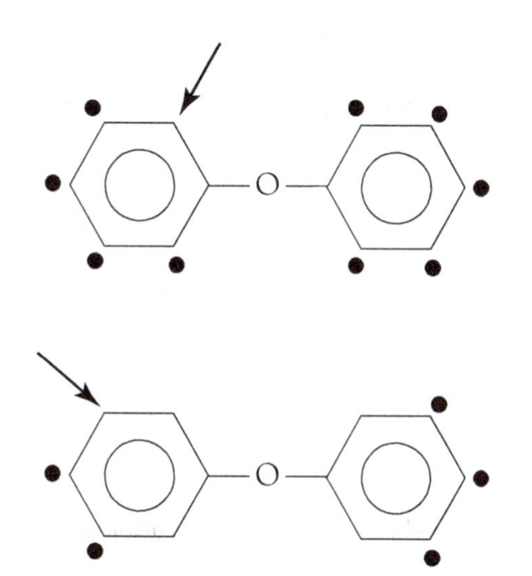

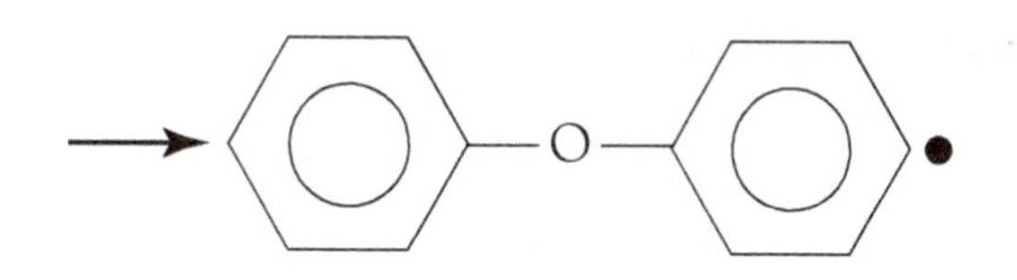

Problem 15–7

We are told concentration $C = Ae^{kt}$

Thus for two different years, t_2 and t_1, we can take the ratio of concentration and thereby eliminate A as an unknown:

$$C_2 / C_1 = e^{kt_2} / e^{kt_1}$$

Taking the natural logarithm of both sides, then

$$\ln (C_2 / C_1) = kt_2 - kt_1$$

so we can solve for k:

$$k = \{ \ln (C_2 / C_1) \} / (t_2 - t_1)$$

Substituting the data for 2000 (t_2) and 1990 (t_1), we obtain

$$k = \{ \ln (7000 / 1100) \} / (2000 - 1990)$$

so $k = \ln (6.364) / 10$

or $k = 0.185$

Hence the doubling time is $0.69 / k = 0.69 / 0.185 = 3.7$ years.

The value of the concentration at any other time t_3 can be obtained by ratio:

$C_3 / C_1 = e^{kt_3} / e^{kt_1}$

so $C_3 = C_1 \, e^{k(t_3 - t_1)}$

or $C_3 = 1100 \, e^{0.185 \times (2010\text{-}1990)}$

$= 45{,}000$

Problem 15–8

The structure at the left shows the hexabrominated ring; no attempt is made in the diagram to show the stereochemistry of the ring.

The triene in the diagram at the right has the identical structure except that a double bond is present between each of the three pairs of carbon atoms that were attached to bromine atoms in the hexabrominated structure.

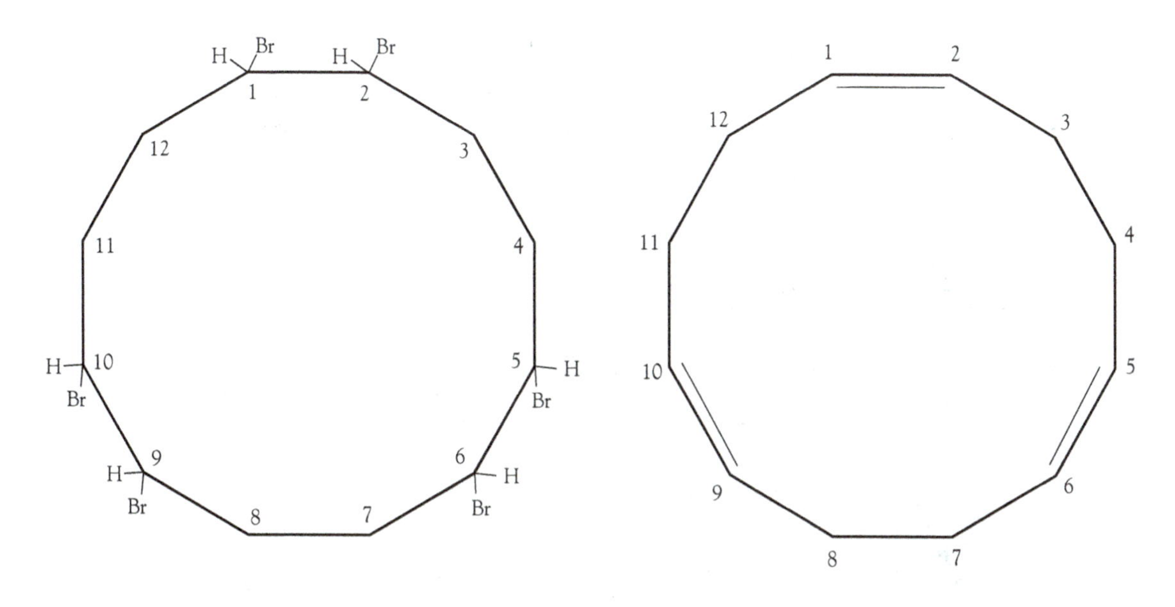

Additional Problems

1. (a) PAH's form from the high-temperature pyrolysis of fat, and form on the surface of foods, rather than in the interior. In the process of barbecuing, fat from the meat drips down onto the hot coals, resulting in significant formation of benzo[a]pyrene, which is carried back up to the meat on the resulting smoke. Such a mechanism cannot occur in oven-broiling, since the heating element is above the meat. In comparing the medium to very well-done burgers, the latter are exposed to the benzo[a]pyrene-containing smoke for a significantly longer period of time.

   (b) 1.52 ng / g burger = 1.52×10^{-9} g / g burger

   Multiplying numerator and denominator of this fraction by 10^9, we obtain 1.52 g / 10^9 g burger = 1.52 ppb

   (c) *Quarter pounder:* 1 lb = 454 g, $\therefore$ pre-cooked weight = 454 g / 4 = 113.5 g

   based on pre-cooked weight: 113.5 g $\times$ 1.52 ng / g $\times$ 0.001 μg / ng = 0.172 μg

2. If free rotation occurs about the C–O bonds, then when the first bromine is missing from an ortho or meta position on the left-side ring, separate isomers corresponding to ortho and meta positions on opposite sides of the right-side ring interchange and cannot both be isolated. The pairs of structures from Problem 15-6b that interconvert when rotation occurs are shown below:

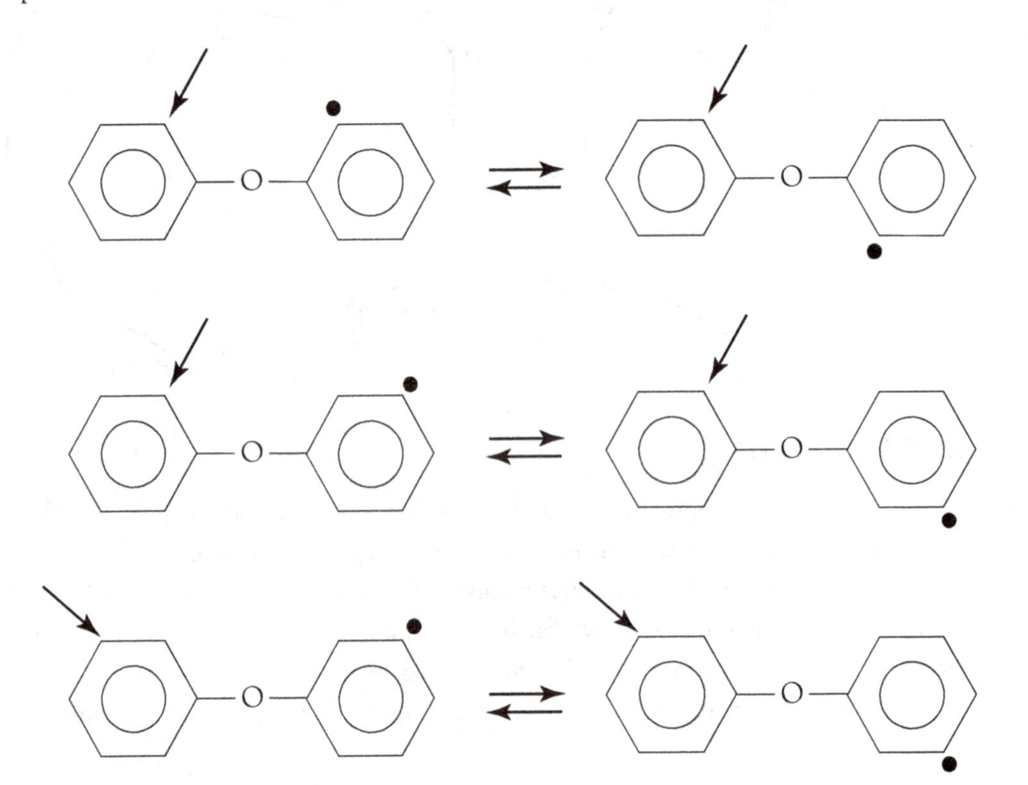

CHAPTER 16

Wastes, Soils, and Sediments

Problem 16-1

(a) The unbalanced equation is

$$CH_2O \longrightarrow CO_2 + CH_4$$

This is readily balanced

$$2\,CH_2O \longrightarrow CO_2 + CH_4$$

(b) The mass of garbage that can biodegrade is 20% of 1 kilogram, or 200 grams. It if decomposes evenly over 20 years, then in one year 10 grams of it decomposes. To calculate the volume of methane it releases, we first determine the moles of CH_2O and convert to moles CH_4:

$$10\text{ g } CH_2O \times \frac{1\text{ mole } CH_2O}{30.03\text{ g } CH_2O} \times \frac{1\text{ mole } CH_4}{2\text{ moles } CH_2O}$$

$$= 0.167\text{ moles } CH_4$$

We use the ideal gas law to determine the volume at 15°C (= 288 K) and 1.0 atmospheric pressure:

$$PV = nRT$$

$$\text{so } V = nRT / P = 0.167\text{ mol} \times 0.082\text{ L atm mol}^{-1}\text{ K}^{-1} \times 288\text{ K} / 1.0\text{ atm}$$

$$= 3.9\text{ L}$$

Thus each kilogram of garbage yields 3.9 L of methane annually.

Problem 16-2

The formation reaction for Fe_2O_3 is

$$2\,Fe(s) + 3/2\,O_2\,(g) \longrightarrow Fe_2O_3(s)$$

The formation of iron from Fe_2O_3 is the reverse of this reaction, so it is endothermic by 824 kJ per 2 moles of Fe. We can convert to a given basis for iron by using the molar mass:

$$\frac{824 \text{ kJ}}{2 \text{ moles Fe}} \times \frac{1 \text{ mole Fe}}{55.85 \text{ g Fe}} = 7.38 \text{ kJ/g}$$

Since the energy saved by not needing to reform aluminum metal from its ore is 31 kJ per gram, whereas for iron it is 7 kJ per gram, recyclers pay much less per kilogram for scrap iron.

Problem 16-3

To deduce formulas from percent composition, recall from your year one chemistry course that you take 100 grams of the compound and first deduce the number of moles of each element in it:

$$50.7 \text{ g C} \times \frac{1 \text{ mole C}}{12.01 \text{ g C}} = 4.225$$

$$4.27 \text{ g H} \times \frac{1 \text{ mole H}}{1.01 \text{ g H}} = 4.228$$

$$45.1 \text{ g O} \times \frac{1 \text{ mole O}}{16.0 \text{ g O}} = 2.819$$

Thus the ratio of C:H:O is 4.22:4.228:2.819, or 1.497:1.500:1. Multiplying by 2 gives a simple ratio of numbers, all of which are close to integers here:

2.995:3.00:2 or 3:3:2

Thus the simplest empirical formula is $C_3H_3O_2$.

Problem 16-4

$$20 \text{ centimoles} \times \frac{0.01 \text{ moles}}{1 \text{ centimole}} \times \frac{1000 \text{ millimoles}}{1 \text{ mole}} = 200 \text{ millimoles}$$

Thus there are 200 millimoles per 1000 grams, or 20 millimoles per 100 grams. Thus the CEC value in both units is the same!

Problem 16-5

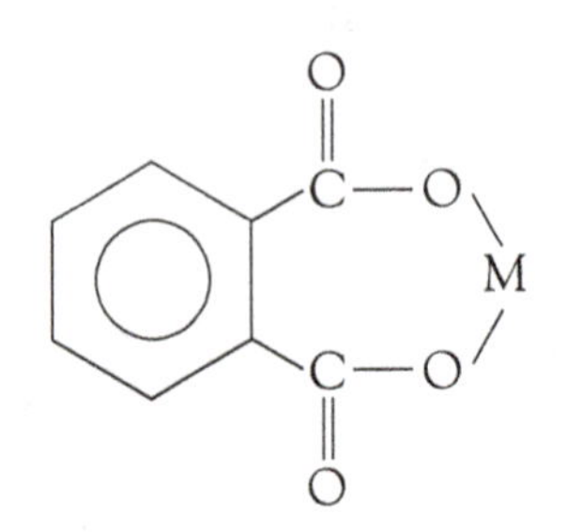

Problem 16-6

The PCB of interest is

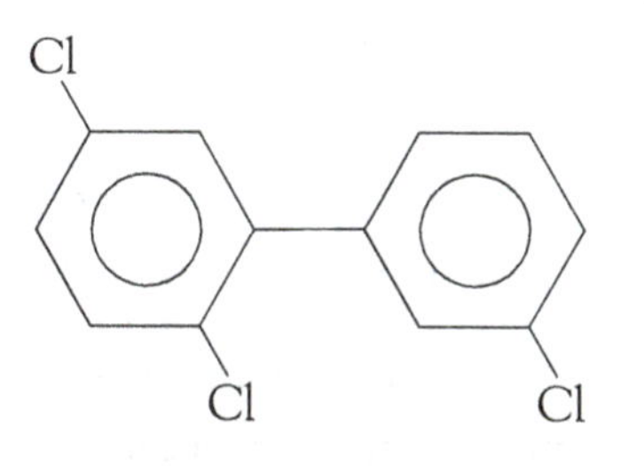

The ortho, meta pair of nonsubstituted sites is available only in the ring at the right, so 2,3 hydroxylation will occur there, followed by splitting of its 1,2 CC bonds. Thus the remaining benzene ring is that shown above at the left, with chlorines at positions ortho and meta to the COOH group.

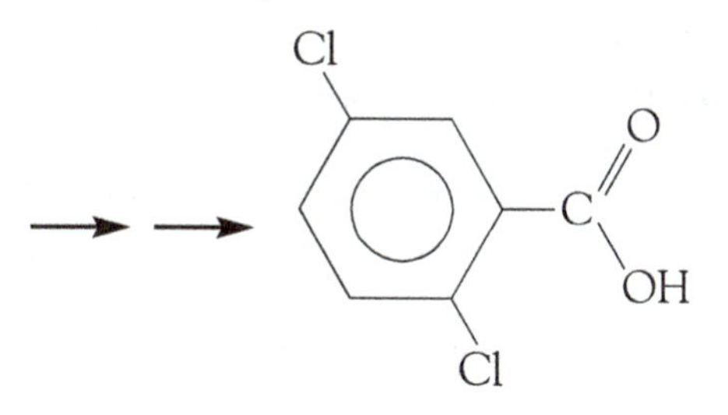

2,5-dichlorobenzoic acid

Problem 16-7

(a) The unbalanced equation is

$$C_{12}H_6Cl_4 + O_2 \longrightarrow CO_2 + H_2O + HCl$$

Balancing the Cl, C, H, and then O in turn yields

$$2\,C_{12}H_6Cl_4 + 25\,O_2 \longrightarrow 24\,CO_2 + 2\,H_2O + 8\,HCl$$

(b) The unbalanced equation is

$$C_{12}H_6Cl_4 + H_2 \longrightarrow CH_4 + HCl$$

Balancing the Cl, C, and then, after noting there are 6 H in $C_{12}H_6Cl_4$, H_2 in turn yields

$$C_{12}H_6Cl_4 + 23\,H_2 \longrightarrow 12\,CH_4 + 4\,HCl$$

Green Chemistry Problems

1. (a) 1. The use of alternative synthetic pathways.

   (b) 1. The prevention of waste.

   6. Energy reduction.

   7. A raw material feedstock should be renewable rather than depleting whenever technically and economically practical.

2. Prevention of waste, use of renewable feedstocks, and lower energy use.

3. (a) 1. Alternative synthetic pathways

   (b) 1. Prevention of waste.

   5. The use of auxiliary substances (e.g., solvents, separation agents, etc.) should be made unnecessary whenever possible and innocuous when used.

   6. Energy requirements should be recognized for their environmental and economic impacts and should be minimized. Synthetic methods should be conducted at ambient temperature and pressure.

4. • Closed loop recyclable.

   • Low toxicity; eliminates the environmental and health concerns of vinyl chloride monomer, phthalates, and combustion products associated with PVC.

   • Recycling requires no additional petroleum feedstocks, which not only benefits the environment but also the economic bottom line.

   • Recycling reduces the amount of landfill space needed.

   • The polyolefin-backed carpeting is 40% lighter in weight than carpeting backed with PVC. This results in the fact that more carpeting can be shipped in the same truck without surpassing weight limits, thus lowering fuel consumption, cost, and pollution.

   • The use of polyolefins eliminates the need for the energy-intensive heating process that is required for PVC. Again, this results in lowering the fuel consumption, cost, and pollution.

   • For carpet backing (including PVC backing) significant amounts of inorganic fillers are used to provide loft and bulk. Traditionally virgin calcium carbonate has been employed for this purpose. EcoWorx contains 60% class C fly ash (a waste byproduct from the burning of lignite or subbituminous coal) as a filler. Using fly ash as a filler thus utilizes an unwanted byproduct and precludes the use of a virgin chemical.

Additional Problems

1. Of the 2000 grams of MSW per day per person, one-quarter or 500 grams of it decomposes anaerobically. Since in this process methane and carbon dioxide are produced approximately in equal molar amounts, and since methane's molar mass is (16.04/60.05) of their total molar mass, the mass of methane in the 500 gram mixture is $(16.04/60.05) \times 500 = 133$ grams. The number of moles of methane equivalent to this mass is $133 / 16.04 = 8.33$ moles. Since from Chapter 6 the heat released per mole of methane combusted is 890 kJ, then burning this quantity of methane liberates $8.33 \times 890 = 7.4 \times 10^3$ kJ $= 7.4 \times 10^6$ J of heat. Thus the total heat liberated from the garbage input of all residents is $300{,}000 \times 7.4 \times 10^3$ kJ $= 2.2 \times 10^9$ kJ. Thus per year the total heat evolved is 365 times this value, or 8×10^{11} kJ. Thus the number of homes, each requiring 1×10^8 kJ, that could be heated is $(8 \times 10^{11}) / (1 \times 10^8) =$

8000 homes. Notice in the calculations above that we have assumed that the MSW input has been occurring sufficiently long that a steady state has been built up, with as much new MSW being added in a day as old MSW decomposes.

2. According to solubility rules, the only common sulfides that are freely soluble—and thus would not have their sediment availabilities determined by AVS—are Na_2S, K_2S, $(NH_4)_2S$, MgS, CaS, SrS, and BaS. Of the metals listed, the doubly charged ones form insoluble carbonates, i.e., $MgCO_3$, $CaCO_3$, $SrCO_3$, and $BaCO_3$, so these metals would be unavailable.

3. First convert the Fe^{2+} concentration into moles and then into moles of sulfide to see how much of the AVS is consumed by it:

$$450 \times 10^{-6}\text{g FeS} \times \frac{1 \text{ mole FeS}}{87.98 \text{ FeS}} = 5.12 \times 10^{-6} \text{ moles FeS} = 5.12 \text{ micromoles FeS}$$

Since 1 mole of FeS requires 1 mole of sulfide, the amount remaining to complex with other metals is $10 - 5.12 = 4.88$ micromoles. Thus since sulfide forms HgS, the moles of Hg^{2+} per gram of soil that can be tied up is also 4.88 micromoles. We can convert this amount into a mass of mercury per gram of soil, then to a tonne of soil:

$$\frac{4.88 \times 10^{-6}\text{moles HgS}}{1 \text{ g soil}} \times \frac{232.65 \text{ g HgS}}{1 \text{ mole HgS}} \times 1000 \text{ kg soil} \times \frac{1000 \text{ g soil}}{1 \text{ kg soil}}$$
$$= 1.14 \times 10^{3}\text{grams Hg} = 1.14 \text{ kg Hg}$$

4. The PCB in question is

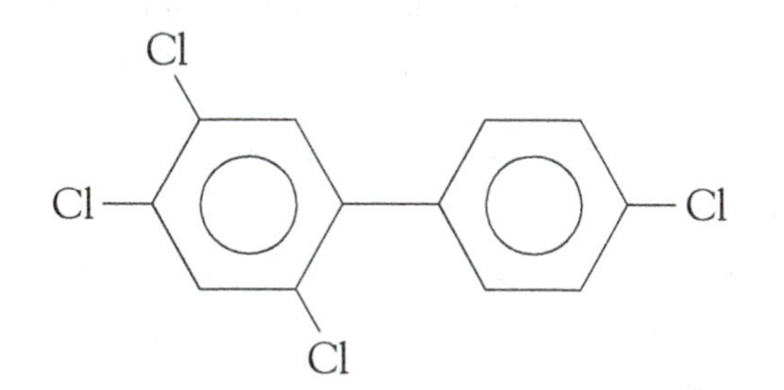

(a) Since only the ring on the right side (above diagram) has both 2 and 3 positions unsubstituted by chlorine, it will undergo reactions leading to its decomposition. Eventually the 1′ carbon will form the —COOH unit, so the acid that forms is

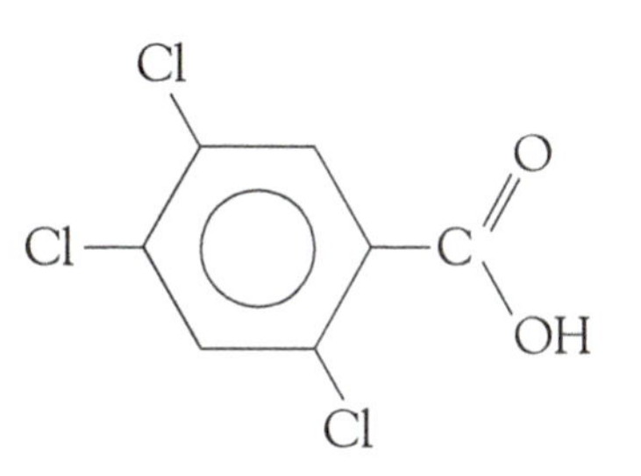

2, 4, 5 - trichlorobenzoic acid

(b) Anaerobic dechlorination of PCBs preferentially involves meta and para chlorines, so the 2 position will remain unchanged. The PCB with only one chlorine remaining therefore is 2-chlorobiphenyl. If two chlorines remain, any one of the other three are possible because they all are either meta or para chlorines. Thus the three dichlorobiphenyls that could result are the 2, 4 -, the 2, 5 -, and the 2, 4′ congeners.

6. The balanced equation for glucose fermentation to ethanol is:

$$C_6H_{12}O_6\ (aq) \longrightarrow 2\ CH_3CH_2OH + 2\ CO_2\ (g)$$

Conversion of 1.0 tonne hardwood scraps:

1.0 tonne $\times$ 1000 kg / tonne $\times$ 1000 g / kg $\times$ 0.46 = 4.6×10^5 g cellulose

mol glucose produced = 4.6×10^5 g / 162.16 g mol^{-1} = 2.84×10^3 mol

mol ethanol = 2 $\times$ mol glucose = 5.68×10^3 mol

mass ethanol = 5.68×10^3 mol $\times$ 46.08 g mol^{-1} = 2.62×10^5 g

volume ethanol = 2.62×10^5 g / 0.789 g mL^{-1} $\times$ 1L / 1000 mL = 332 L

7. The best range of wavelengths to be absorbed would be the UV-B (280–320 nm). UV-C in sunlight is filtered out by the atmosphere (O_2 and O_3), so would not be useful. However, a significant portion of UV-B does make it to the surface. The intensity of visible light from sunlight is much higher than UV-B, but plastics that are photolyzed by visible light would break down during their indoor use, under exposure to fluorescent and incandescent light. UV-A could potentially be used, but this would still result in plastics that would be less stable during their use, since there is a small but significant amount of UV-A in fluorescent and incandescent light.

The major limitation to the breakdown of these plastics is that they need to be exposed to sunlight, i.e., they need to be on the surface. Such plastics in landfills would soon be covered by other waste and/or topsoil, and thus the photolytic degradation would cease. It would be a useful process for litter thrown along the side of highways, however.

The maximum energy in kJ / mol for a bond that can be cleaved by 300 nm light can be calculated from the equation in Chapter 1, namely:

Energy = 119627 / λ = 119627 / 300 = 399 kJ / mol

CHAPTER 17

The Detailed Free-Radical Chemistry of the Atmosphere

Problem 17-1

CF_2Cl_2 and other CFCs do not contain hydrogen or multiple bonds, so they are not attacked by OH. They also do not absorb light of wavelength regions that reach the troposphere; therefore, photochemical processes do not initiate CFC oxidation either. In contrast, CH_2Cl_2 contains hydrogen, and so it will be attacked by OH and its oxidation initiated:

$$CH_2Cl_2 + OH^\bullet \longrightarrow CHCl_2^\bullet + H_2O$$

Problem 17-2

If the reaction is endothermic it will have a substantial activation energy and therefore it will be slow.

Problem 17-3

Since the reaction is not fast enough to be observed, presumably its activation energy must be large. Thus, the reaction probably is quite endothermic, because most free radical reactions with large activation energies are quite endothermic.

Problem 17-4

H_2 will react with $OH^\bullet$ because it does not absorb visible or UV-A light; because H_2 contains only a single bond, the $OH^\bullet$ abstracts one hydrogen:

$$OH^\bullet + H_2 \longrightarrow H_2O + H^\bullet$$

Since $H^\bullet$ is a radical, it reacts with O_2 by addition to it:

$$H^\bullet + O_2 \longrightarrow HOO^\bullet$$

Peroxy radicals react with $NO^\bullet$, so

$$HOO^\bullet + NO^\bullet \longrightarrow OH^\bullet + NO_2^\bullet$$

Adding up these three reactions, we find the overall reaction involving H_2 to be

$$H_2 + O_2 + NO^{\bullet} \longrightarrow H_2O + NO_2^{\bullet}$$

Problem 17-5

Since methanol contains no multiple bonds and does not absorb visible or UV-A light, it will react with $OH^{\bullet}$ by hydrogen atom abstraction. First, consider that it is a hydrogen bonded to carbon that is abstracted:

$$H_3COH + OH^{\bullet} \longrightarrow H_2\dot{C}OH + H_2O$$

The carbon-centered radical reacts with O_2; because by abstracting the H bonded to O, a stable molecule forms and the C—O bond is converted to C ═ O, this process rather than O_2 addition subsequently occurs:

$$H_2\dot{C}OH + O_2 \longrightarrow H_2CO + HOO^{\bullet}$$

In an alternative scenario, the $OH^{\bullet}$ abstracts the hydrogen bonded to oxygen from the original methanol molecule:

$$H_3COH + OH^{\bullet} \longrightarrow H_3\dot{C}O + H_2O$$

In this case, the O_2 molecule attacks and abstracts a hydrogen bonded to carbon because formaldehyde, with its C ═ O bond, is thereby formed.

$$H_3\dot{C}O + O_2 \longrightarrow H_2CO + HOO^{\bullet}$$

Problem 17-6

Carbon monoxide is a non-radical and does not absorb visible or UV-A light; therefore, it must react with $OH^{\bullet}$. Since it contains a C ≡ O bond, the $OH^{\bullet}$ adds to the CO molecule:

$$CO + OH^{\bullet} \longrightarrow H{-}O{-}\dot{C}{=}O$$

The radical reacts with O_2, and does so by giving up its hydrogen because thereby the O—C bond is converted to O ═ C

$$H{-}O{-}\dot{C}{=}O + O_2 \longrightarrow HOO^{\bullet} + O{=}C{=}O$$

Finally, because it is a peroxy radical, $HOO^{\bullet}$ is returned to $OH^{\bullet}$ by reaction with $NO^{\bullet}$:

$$HOO^{\bullet} + NO^{\bullet} \longrightarrow OH^{\bullet} + NO_2^{\bullet}$$

Adding the three steps and cancelling terms that appear in both sides, we obtain:

$$CO + O_2 + NO^{\bullet} \longrightarrow CO_2 + NO_2^{\bullet}$$

as the overall reaction.

Problem 17-7

The reaction sequence begins with an H abstraction:

$$OH^{\bullet} + H_2CO \longrightarrow H_2O + H{-}\dot{C}{=}O$$

The radical HCO reacts with O_2 also by H abstraction, because this produces the stable molecule CO with a triple bond:

$$H{-}\dot{C}{=}O + O_2 \longrightarrow C{\equiv}O + HOO^{\bullet}$$

As in Additional Problem 17-6, CO is then attacked by $OH^{\bullet}$—it is a stable molecule with a multiple bond, and therefore the $OH^{\bullet}$ adds to it:

$$OH^{\bullet} + CO \longrightarrow H{-}O{-}\dot{C}{=}O$$

Finally, O_2 abstracts the H from HOCO because the stable molecule CO_2 is thereby formed:

$$H{-}O{-}\dot{C}{=}O + O_2 \longrightarrow HOO^{\bullet} + O{=}C{=}O$$

The two $HOO^{\bullet}$ return to $OH^{\bullet}$ by reaction with $NO^{\bullet}$

$$2\,[HOO^{\bullet} + NO^{\bullet} \longrightarrow OH^{\bullet} + NO_2^{\bullet}]$$

Summing the reactions, we obtain

$$H_2CO + 2\,NO^{\bullet} + 2\,O_2 \longrightarrow CO_2 + H_2O + 2\,NO_2^{\bullet}$$

There is no increase in the number of free radicals; an increase would occur only when photochemical decomposition of a non-radical into two radicals is involved.

Problem 17-8

Substituting H for R in the net reaction given in the text, we obtain:

$$H_2C = CH_2 + OH^{\bullet} + 2\,O_2 + NO^{\bullet} \longrightarrow 2\,H_2CO + HOO^{\bullet} + NO_2^{\bullet}$$

The formaldehyde molecules undergo photochemical decomposition to give $H^{\bullet} + HCO^{\bullet}$:

$$H_2CO \xrightarrow{UV} H^{\bullet} + HCO^{\bullet}$$

We know from our principles that O_2 will immediately add to $H^{\bullet}$, and the O_2 will abstract the hydrogen from $HCO^{\bullet}$ because the double bond in $H{-}\dot{C}{=}O$ is thereby converted to a triple one in CO:

$$H^{\bullet} + O_2 \longrightarrow HOO^{\bullet}$$
$$HCO^{\bullet} + O_2 \longrightarrow HOO^{\bullet} + CO$$

As previously discussed, the CO adds OH and the resulting HOCO reacts with O_2 by hydrogen abstraction because the number of CO bonds is thereby increased:

$$CO + OH^\bullet \longrightarrow HOCO^\bullet$$
$$HOCO^\bullet + O_2 \longrightarrow HOO^\bullet + CO_2$$

When twice these reactions (because 2 H_2CO must be destroyed) are added to the net reaction above, and after cancellation of common terms, we obtain:

$$H_2C{=}CH_2 + 8\,O_2 + NO^\bullet + 3\,OH^\bullet \xrightarrow{UV} 2\,CO_2 + NO_2^\bullet + 7\,HOO^\bullet$$

The $HOO^\bullet$ oxidation of $NO^\bullet$ involves 7 molecules of each:

$$7\,[HOO^\bullet + NO^\bullet \longrightarrow OH^\bullet + NO_2^\bullet]$$

Thus the final net reaction is:

$$H_2C{=}CH_2 + 8\,O_2 + 8\,NO^\bullet \xrightarrow{UV} 2\,CO_2 + 8\,NO_2^\bullet + 4\,OH^\bullet$$

Problem 17-9

$$CH_3(H)CO \xrightarrow{sunlight} CH_3^\bullet + HCO^\bullet$$

$$HCO^\bullet + O_2 \longrightarrow HOO^\bullet + CO$$
$$CO + OH^\bullet \longrightarrow HOCO^\bullet$$
$$HOCO^\bullet + O_2 \longrightarrow CO_2 + HOO^\bullet$$

$$CH_3^\bullet + O_2 \longrightarrow CH_3OO^\bullet$$
$$CH_3OO^\bullet + NO^\bullet \longrightarrow CH_3O^\bullet + NO_2^\bullet$$
$$CH_3O^\bullet + O_2 \longrightarrow H_2CO + HOO^\bullet$$

$$H_2CO \xrightarrow{light} H^\bullet + HCO^\bullet$$
$$H^\bullet + O_2 \longrightarrow HOO^\bullet$$
$$HCO^\bullet + O_2 \longrightarrow HOO^\bullet + CO$$
$$CO + OH^\bullet \longrightarrow HOCO^\bullet$$
$$HOCO^\bullet + O_2 \longrightarrow CO_2 + HOO^\bullet$$

$$6\,[HOO^\bullet + NO^\bullet \longrightarrow OH^\bullet + NO_2^\bullet]$$

Overall, the reaction is $CH_3(H)CO + 7\,O_2 + 7\,NO^\bullet \longrightarrow 2\,CO_2 + 7\,NO_2^\bullet + 4\,OH^\bullet$

Problem 17-10

From the text, the dominant reaction at 5–8 AM is the production of NO_2 from NO, and that from 8 AM–12 noon is the production of aldehydes from hydrocarbons.

Problem 17-11

The initial reaction must be

$$H_2CO \xrightarrow{\text{sunlight}} H_2 + CO$$

Since H_2 contains H, we expect $OH^\bullet$ to abstract a hydrogen from it:

$$OH^\bullet + H_2 \longrightarrow H_2O + H^\bullet$$

We expect $H^\bullet$ will add O_2

$$H^\bullet + O_2 \longrightarrow HOO^\bullet$$

The $HOO^\bullet$ will oxidize $NO^\bullet$

$$HOO^\bullet + NO^\bullet \longrightarrow OH^\bullet + NO_2^\bullet$$

From previous examples, we know $OH^\bullet$ adds to CO and then O_2 abstracts hydrogen from the radical:

$$OH^\bullet + CO \longrightarrow HOCO^\bullet$$
$$HOCO^\bullet + O_2 \longrightarrow HOO^\bullet + CO_2$$
$$HOO^\bullet + NO^\bullet \longrightarrow OH^\bullet + NO_2^\bullet$$

Adding together all six steps involving radicals, we obtain

$$H_2 + CO + 2\,O_2 + 2\,NO^\bullet \longrightarrow H_2O + CO_2 + 2\,NO_2^\bullet$$

Adding this to the first step, we obtain:

$$H_2CO + 2\,O_2 + 2\,NO^\bullet \xrightarrow{\text{sunlight}} H_2O + CO_2 + 2\,NO_2^\bullet$$

Problem 17-12

$$H_2C{=}CH_2$$
$$\downarrow OH^\bullet$$
$$H_2\dot{C}{-}CH_2OH$$
$$\downarrow O_2$$
$$H_2C(O{-}O^\bullet){-}CH_2OH$$
$$\downarrow NO^\bullet$$
$$H_2C(O^\bullet){-}CH_2OH$$
$$\downarrow \text{cleaves (see text)}$$
$$H_2CO + \dot{C}H_2OH$$
$$\downarrow O_2$$
$$H_2CO + HOO^\bullet$$

Both H_2CO

$$\downarrow \text{light}$$
$$H^\bullet + HCO^\bullet$$
$$\downarrow O_2 \quad \downarrow O_2$$
$$HOO^\bullet \quad HOO^\bullet + CO$$
$$\downarrow OH^\bullet$$
$$HO\dot{C}O$$
$$\downarrow O_2$$
$$CO_2 + HOO^\bullet$$

In all instances, $HOO^\bullet + NO^\bullet \longrightarrow OH^\bullet + NO_2^\bullet$

Problem 17-13

If $CH_3OO^\bullet$ combines with $HOO^\bullet$, we have a four-oxygen chain that presumably will split out at least one O_2 molecule.

$$CH_3OO^\bullet + HOO^\bullet \longrightarrow [CH_3OOOOH] \longrightarrow O_2 + CH_3OOH$$

Problem 17-14

Items b, c, and d do not contain "loose" oxygens.

Problem 17-15

From the Systematics section we find that

(a) NO abstracts an oxygen atom from all species containing a loose oxygen, i.e., O_3, ClO, BrO, and HOO.

(b) O abstracts an oxygen from all species with "loose" oxygens, i.e., from O_3, ClO, BrO, HOO, and NO_2.

(c) Sunlight is absorbed in the UV region by all the species O_3, ClO, BrO, HOO, and NO_2, and an oxygen atom is consequently detached.

(d) YOOY molecules of the type ClOOCl and $(NO_2)_2$ and perhaps BrOOBr will form.

(e) O_2 is produced when two YO molecules react but result in a chain that has more than two O atoms bonded to each other; thus O_2 is obtained when 2 HOO react and when two O_3 react.

Problem 17-16

(a) $BrO^\bullet + O \longrightarrow Br^\bullet + O_2$

(b) $BrO^\bullet + ClO^\bullet \longrightarrow Br^\bullet + Cl^\bullet + O_2$

(c) $2\ BrO^\bullet \longrightarrow BrOOBr$

(d) $BrO^\bullet + UV \longrightarrow Br^\bullet + O$

Problem 17-17

(a) $ClO^\bullet + NO_2^\bullet \longrightarrow ClONO_2$

(b) $2\ ClO^\bullet \longrightarrow ClOOCl$

(c) $ClO^\bullet + UV \longrightarrow Cl^\bullet + O$
or $ClOOCl + UV \longrightarrow Cl^\bullet + ClOO^\bullet$
or $ClO^\bullet + O \longrightarrow Cl^\bullet + O_2$
or $ClO^\bullet + NO^\bullet \longrightarrow Cl^\bullet + NO_2^\bullet$

Problem 17-18

$$:\ddot{\underset{\cdot\cdot}{F}}-\dot{\underset{\cdot\cdot}{O}}:$$

Yes, the oxygen should be "loose" because it is joined by a single bond to an electronegative atom that possesses lone pairs.

Problem 17-19

The expected reaction is abstraction of the loose oxygen from $ClO^{\bullet}$ by $NO^{\bullet}$:

$$ClO^{\bullet} + NO^{\bullet} \longrightarrow Cl^{\bullet} + NO_2^{\bullet}$$

The $NO_2^{\bullet}$ could decompose in sunlight by the loss of atomic oxygen, or react with atomic oxygen, or with another $NO_2^{\bullet}$:

$$NO_2^{\bullet} \xrightarrow{\text{sunlight}} NO^{\bullet} + O$$
$$NO_2^{\bullet} + O \longrightarrow NO^{\bullet} + O_2$$
$$2\,NO_2^{\bullet} \longrightarrow N_2O_4$$

The $Cl^{\bullet}$ could abstract an oxygen from ozone or a hydrogen from methane:

$$Cl^{\bullet} + O_3 \longrightarrow ClO^{\bullet} + O_2$$
$$Cl^{\bullet} + CH_4 \longrightarrow HCl + CH_3^{\bullet}$$

The cycle would destroy the ozone (by $O_3 + O \longrightarrow 2\,O_2$ overall) if the Cl reaction is that between $Cl^{\bullet}$ and O_3 and the $NO_2^{\bullet}$ reacts with O; if, however, the $NO_2^{\bullet}$ decomposes by sunlight, the net reaction is $O_3 \longrightarrow O_2 + O$, which is followed by $O + O_2 \longrightarrow O_3$, and so no ozone is destroyed.

Box 17-1, Problem 1

(a) Only the oxygen here can be the site for the unpaired electron, which is consistent with it forming only one bond rather than its usual two:

$$:\dot{\underset{\cdot\cdot}{O}}-H$$

(b) Only the carbon here can be the site for the unpaired electron, which is consistent with it forming only three bonds rather than its usual four:

$$H_2\dot{C}-H$$

(c) Since F and Cl must form (single) bonds to the carbon, neither can be the site of the unpaired electron, which therefore must be located on the carbon, which forms three rather than four bonds:

$$:\ddot{\underset{..}{F}} \diagdown \dot{C} - \ddot{\underset{..}{Cl}}: \quad (\text{second } :\ddot{\underset{..}{F}} \diagup \text{ on } C)$$

(d) The carbon here must form four bonds, and therefore cannot be the site of the unpaired electron, and similarly for the central oxygen, which must form two bonds. The unpaired electron must therefore be located on the terminal oxygen atom, as in HOO:

$$\mathrm{H}-\overset{\mathrm{H}}{\underset{\mathrm{H}}{\overset{|}{\underset{|}{\mathrm{C}}}}}-\ddot{\underset{..}{\mathrm{O}}}-\dot{\underset{..}{\mathrm{O}}}:$$

(e) As in (d), the unpaired electron must be on the terminal oxygen:

$$\mathrm{H}-\overset{\mathrm{H}}{\underset{\mathrm{H}}{\overset{|}{\underset{|}{\mathrm{C}}}}}-\dot{\underset{..}{\mathrm{O}}}:$$

(f) As in (d), the unpaired electron must be on the terminal oxygen. The Cl atom must form its usual one single bond and thus cannot be the site:

$$:\ddot{\underset{..}{\mathrm{Cl}}}-\ddot{\underset{..}{\mathrm{O}}}-\dot{\underset{..}{\mathrm{O}}}:$$

(g) As in (f), the unpaired electron must be on the terminal oxygen. The Cl atom must form its usual one single bond and thus cannot be the site:

$$:\ddot{\underset{..}{\mathrm{Cl}}}-\dot{\underset{..}{\mathrm{O}}}:$$

(h) The oxygen here cannot be the site of the unpaired electron, since that would require the central carbon atom to form only two bonds (as in H – C – O). Hence the oxygen must form two bonds to carbon, and the hydrogen one bond to C. Thus the site of the unpaired electron here is the carbon atom:

$$\mathrm{H}-\dot{\mathrm{C}}=\ddot{\mathrm{O}}:$$

(i) In the best structure for NO, the oxygen atom forms two bonds to the nitrogen atom, which is the site of the unpaired electron and forms two rather than its usual three bonds:

$$:\dot{\mathrm{N}}=\underset{..}{\mathrm{O}}:$$

Additional Problems

1. Since CO is a stable molecule containing multiple bonds, it reacts with $OH^•$ by addition of the free radical to the C atom (since addition to the O gives a OO bond):

$$CO + OH^• \longrightarrow HOCO^• \quad (\text{i.e., } H\text{-}O\text{-}\dot{C}\text{=}O)$$

The free radical $HOCO^•$ reacts with O_2, by abstraction of H since this allows the addition of a new CO bond, in CO_2:

$$HOCO^• + O_2 \longrightarrow HOO^• + CO_2 \quad (\text{i.e., } O\text{=}C\text{=}O)$$

The HOO radical reacts in the usual way with NO:

$$HOO^• + NO^• \longrightarrow OH^• + NO_2^•$$

If NO_2 photolyzes, it produces atomic O, which then adds to O_2 to produce ozone:

$$NO_2^• \xrightarrow{UV} NO^• + O$$

$$O + O_2 \longrightarrow O_3$$

Adding up all these steps, and cancelling common terms, gives the result

$$CO + 2\,O_2 \xrightarrow{UV} CO_2 + O_3$$

Thus the concentration of ozone is increased by this process involving $OH^•$ as a catalyst.

2. Since $H_3C—CH_3$ has no double bonds, initially $OH^•$ reacts by abstracting a hydrogen from it. Then the carbon-centered radical adds O_2, and the peroxyl radical oxidizes $NO^•$:

$$H_3C—CH_3 \xrightarrow{OH^•} \underset{(+H_2O)}{H_3C—\dot{C}H_2} \xrightarrow{O_2} H_3C—CH_2OO^•$$

$$\xrightarrow{NO^•} \underset{(+NO_2^•)}{H_3C—CH_2O^•}$$

This oxygen-central radical can be transformed to one containing $C{=}O$ if one of the CH_2's H atoms is abstracted by O_2. The resulting aldehyde is then photolyzed:

$$H_3C—CH_2O^• \xrightarrow{O_2} \underset{(+HOO^•)}{H_3C—CH{=}O} \xrightarrow{\text{photon}} H_3C^• + HCO^•$$

The $HCO^•$ radical has its H abstracted by O_2 because a CO triple bound is formed; $OH^•$ then adds to this bond, and oxygen abstracts the oxygen to give the final product, carbon dioxide:

$$HCO^• \xrightarrow{O_2} \underset{(+HOO^•)}{CO} \xrightarrow{OH^•} HOCO^• \xrightarrow{O_2} \underset{(+HOO^•)}{CO_2}$$

Meanwhile the methyl radical $H_3C^\bullet$ adds O_2 to produce a peroxyl radical, which then oxidizes $NO^\bullet$, and has an H abstracted to form a C = O bond:

$$H_3C^\bullet \xrightarrow{O_2} H_3COO^\bullet \xrightarrow{NO^\bullet} \underset{(+NO_2^\bullet)}{H_3CO^\bullet} \xrightarrow{O_2} \underset{(+HOO^\bullet)}{H_2C{=}O}$$

Formaldehyde undergoes photochemical decomposition to $H^\bullet$, which adds O_2, and to $HCO^\bullet$, which reacts as described above:

$$H_2CO \xrightarrow{\text{photon}} H^\bullet + HCO^\bullet \xrightarrow{2\,O_2} HOO^\bullet + HOO^\bullet + CO$$

$$CO \xrightarrow{OH^\bullet} HOCO^\bullet \xrightarrow{O_2} CO_2 + HOO^\bullet$$

Summing up all the reactions, we obtain:

$$H_3CCH_3 + 9\,O_2 + 2\,NO^\bullet + 3\,OH^\bullet \longrightarrow 2\,CO_2 + 2\,NO_2^\bullet + 7\,HOO^\bullet + H_2O$$

Finally, adding in the conversion of 7 $HOO^\bullet$ back to 7 $OH^\bullet$

$$7\,[HOO^\bullet + NO^\bullet \longrightarrow OH^\bullet + NO_2^\bullet]$$

we obtain the overall reaction:

$$H_3CCH_3 + 9\,O_2 + 9\,NO^\bullet \longrightarrow 2\,CO_2 + 9\,NO_2^\bullet + H_2O + 4\,OH^\bullet$$

3. Carbon monoxide reacts with $OH^\bullet$ radical to produce $HOCO^\bullet$, which then reacts with O_2 to produce $HOO^\bullet$ radicals

$$CO \xrightarrow{OH^\bullet} HOCO^\bullet \xrightarrow{O_2} CO_2 + HOO^\bullet$$

If $HOO^\bullet$ reacts with ozone, a chain of five oxygen atoms result, so two O_2's and OH result:

$$HOO^\bullet + O_3 \longrightarrow [HOOOOO]^\bullet \longrightarrow OH^\bullet + 2\,O_2$$

Thus the overall reaction is:

$$CO + O_3 \longrightarrow CO_2 + O_2$$

Thus the concentrations of O_3 should be abnormally low, because it is destroyed by CO.

4. (i) $CH_3CH_2CH_3 + OH^\bullet \longrightarrow CH_3\dot{C}HCH_3 + H_2O$

   (secondary carbon radical more stable than a primary carbon radical)

   (ii) $H_2C{=}CHCH_3 + OH^\bullet \longrightarrow HOCH_2\dot{C}HCH_3$

   (abstraction of an H atom from the $-CH_3$ may also be likely)

   (iii) $HCl + OH^\bullet \longrightarrow$ no reaction

   (very strong HCl bond)

(iv) $H_2O + OH^{\bullet} \longrightarrow OH + H_2O$

(no net reaction!)

5. NO_2: $\cdot\ddot{\underset{\cdot\cdot}{O}}-\ddot{N}=\dot{\underset{\cdot}{O}}\!:$

HNO_2: $H-\ddot{\underset{\cdot\cdot}{O}}-\ddot{N}=\dot{\underset{\cdot}{O}}\!:$

HNO_3: $H-\ddot{\underset{\cdot\cdot}{O}}-N(-\ddot{O}\!:)(=\ddot{O})$

6. The energy of the photon absorbed must be equal to or greater than the energy of the bond to be cleaved. In the case of the longest wavelength of light that will cleave the bond, the energy of a photon with this wavelength will be exactly equal to the bond energy:

$\lambda = 119{,}627 \,/\, \text{bond energy}$

In the case of $BrO^{\bullet}$, the bond energy is 235 kJ mol^{-1}:

$\lambda = 119{,}627 \,/\, 235 = 509$ nm

This is in the visible region, specifically the green part of the spectrum.

Similarly, the following can be calculated:

$HOO^{\bullet}$: 450 nm (visible region—blue)

$ClO^{\bullet}$: 440 nm (visible region—blue)

$NO_2^{\bullet}$: 392 nm (UV-A region)

Online Appendix: Background Organic Chemistry

Problem 1

(a) Pentane is an alkane with five carbons. Since the name is preceded by n -, all the carbons lie in an unbranched sequence. Since each terminal C is connected by only one bond to the rest of the carbon chain, it must have three hydrogens attached. Each internal C is attached to two carbons, and so has two hydrogens:

```
     H   H   H   H   H
     |   |   |   |   |
H  — C — C — C — C — C — H
     |   |   |   |   |
     H   H   H   H   H
```

$CH_3\,CH_2\,CH_2\,CH_2\,CH_3$

(b) Hexane has six C's in a chain; ethyl is the group derived from ethane so is has two carbons. Hence, the carbon backbone is:

```
C — C — C — C — C — C
        |
        C
        |
        C
```

Given the number of C — C bonds formed by each carbon, it follows that the number of H atoms and the complete formulae are:

```
     H   H   H   H   H   H
     |   |   |   |   |   |
H  — C — C — C — C — C — C — H
     |   |   |   |   |   |
     H   H   C   H   H   H
           /  |  \
          H   |   H
           H— C —H
              |
              H
```

$(CH_3\,CH_2)_2\,CH\,CH_2\,CH_2\,CH_3$

(c) The carbon backbone must be as follows, since methyl has one carbon and butane has four:

```
    C   C
    |   |
C — C — C — C
```

Thus, the full formulae are:

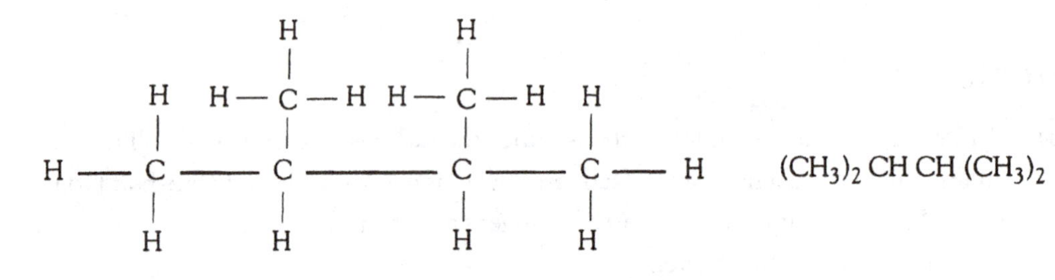

Problem 2

(a) Propene has the backbone $\overset{①}{C} = \overset{②}{C} - \overset{③}{C}$

so the full formula is:

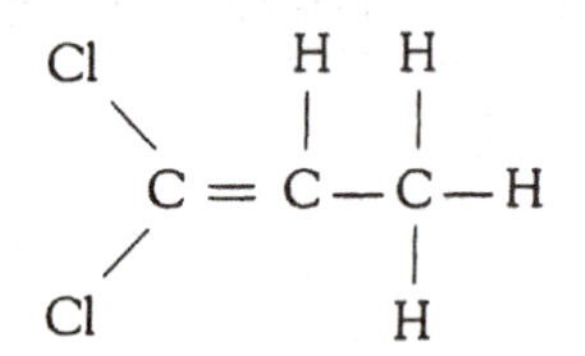

The number of H's at C_2 and C_3 are determined by the condition that the number of bonds formed by each carbon is four.

(b) "Per" means all possible positions of substitution are occupied by the same group, here chlorine:

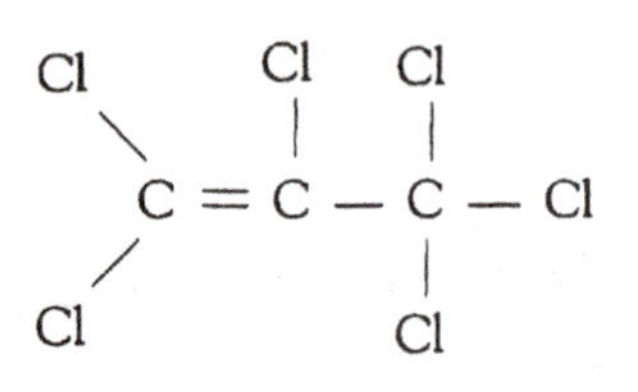

(c) 2- butene must have four carbons and one double bond at the 2, 3 position, since the prefix is 2; its skeleton is:

$$C - C = C - C$$

The terminal C's have three H's each, and the central C's have one H each to make a total of four bonds for each carbon:

```
      H    H    H    H
      |    |    |    |
H  —  C  — C  = C  — C  — H
      |              |
      H              H
```

Problem 3

(a) There are only two carbons in the molecule and they form a single bond to each other, so the molecule is an ethane derivative:

1, 1, 2, 2 - tetrachloroethane

(b) The carbon network has four atoms and one double bond, so it is a butene. We number it starting from the right side so as to give the double bond the lowest possible number:

1 - butene

(c) The molecule has four C's and two C $=$ C, so it is a butadiene. We display the number of the first C's in each double bond:

```
①   ②   ③   ④
C = C — C = C
```

1, 3 - butadiene

Problem 4

(a) The carbon backbone must be:

```
C — C — C — C — C
        |
        C
```

Adding hydrogens so each C forms four bonds, we obtain:

```
      H    H          H          H    H
      |    |          |          |    |
H  —  C  — C  ——————  C  ——————  C  — C  — H
      |    |          |          |    |
      H    H     H —  C  — H     H    H
                      |
                      H
```

(b) The carbon backbone must be:

```
C — C — C — C
        ||
        C
```

so the whole structure is:

```
    H   H       H
    |   |       |
H — C — C — C — C — H
    |   |   ||  |
    H   H   C   H
           / \
          H   H
```

(c) The backbone, including the chlorine atom, is:

$$C=C-C=C-C=C-Cl$$

so the whole structure is:

```
H        H   H   H   H   H
 \       |   |   |   |   |
  C  =   C — C = C — C = C — Cl
 /
H
```

Problem 5

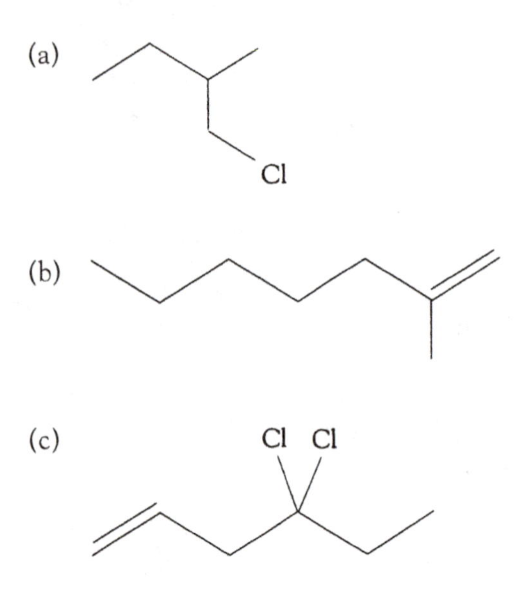

Problem 6

(a) Since ethyl implies two carbons, the framework is

C — C — O — H

Thus, the full structural formula is:

```
    H    H
    |    |
H — C  — C — O — H
    |    |
    H    H
```

and the symbolic diagram is

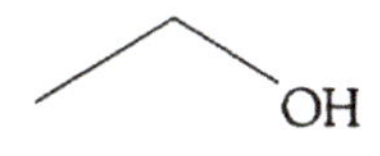

(b) Amine implies —N— (with a third bond drawn above the N); since only one carbon chain is attached to it, it must be an —NH_2 unit:

C—C—NH_2

Thus, the full structural formula is:

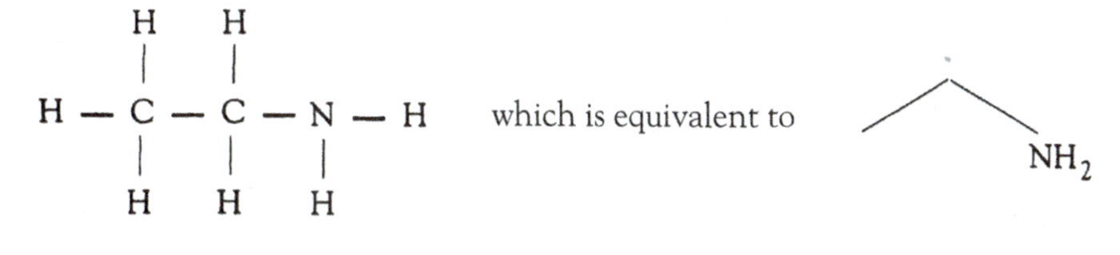

(c) From the text, we know acetic acid is

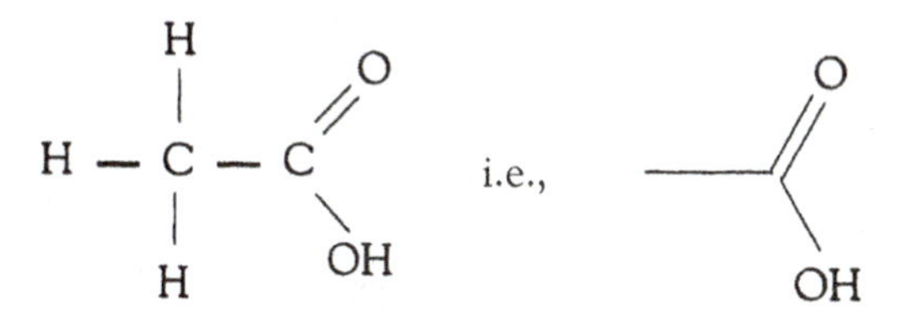

Problem 7

(a) Propane has three C's, so cyclopropane is a ring with three carbons:

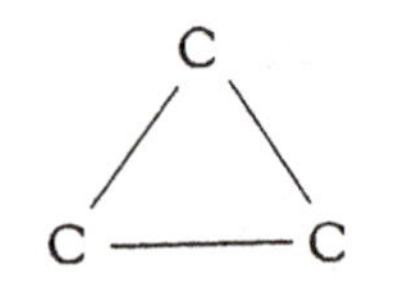

Since each C forms two C — C bonds, each must have two hydrogens:

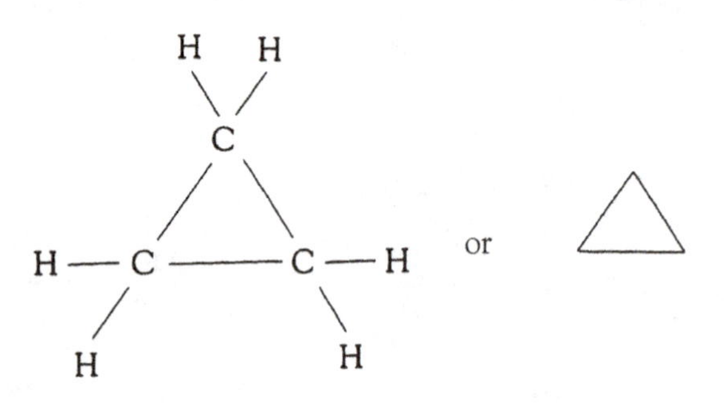

(b) Butane has four C's, so cyclobutane is a ring with four carbons, one of which is bonded to a chlorine:

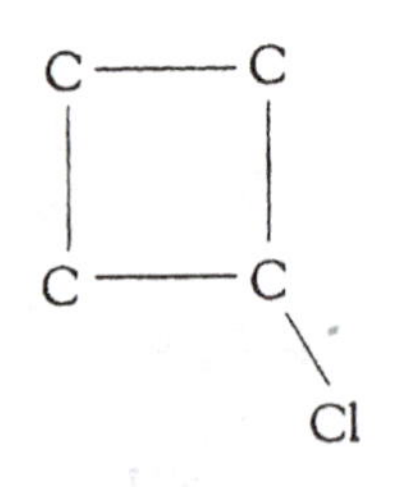

With hydrogens, the formula is

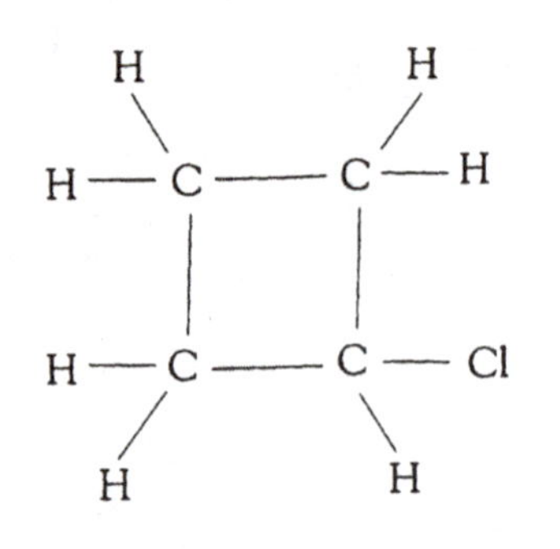

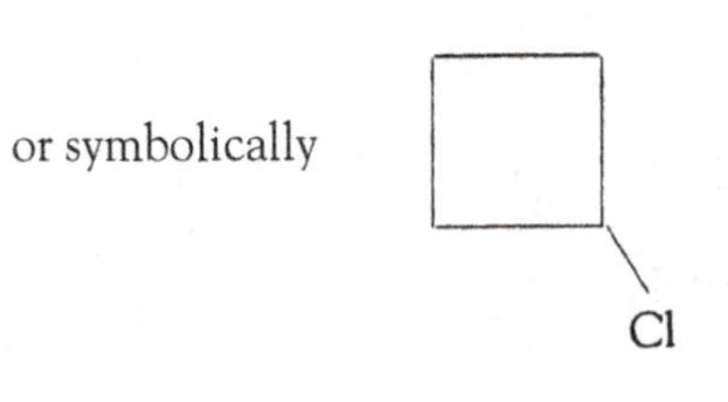

(c) Hexane has 6 carbons, so cyclohexane is a ring of six carbons:

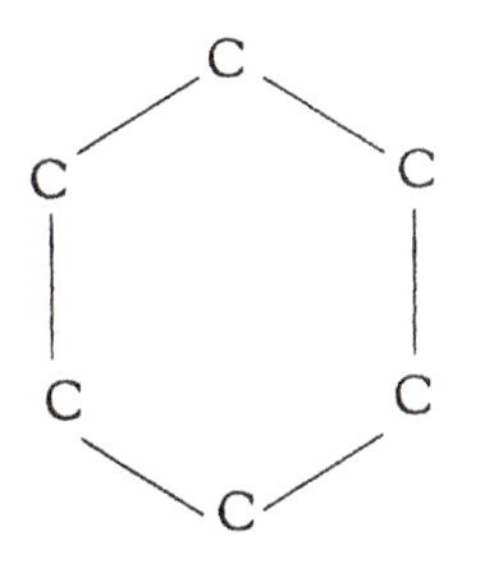

There are several dimethyl isomers, with and without both CH_3's on the same carbon: e.g.,

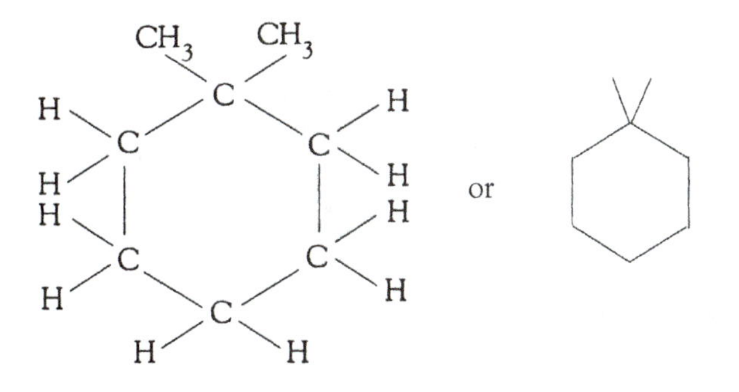

Problem 8

The unique isomers can be generated by first considering that all three Cl's are on adjacent carbons (only one isomer). The next possibility is two adjacent Cl's — then there is only one choice for the position of the third. In the final structure, no Cl's are adjacent.

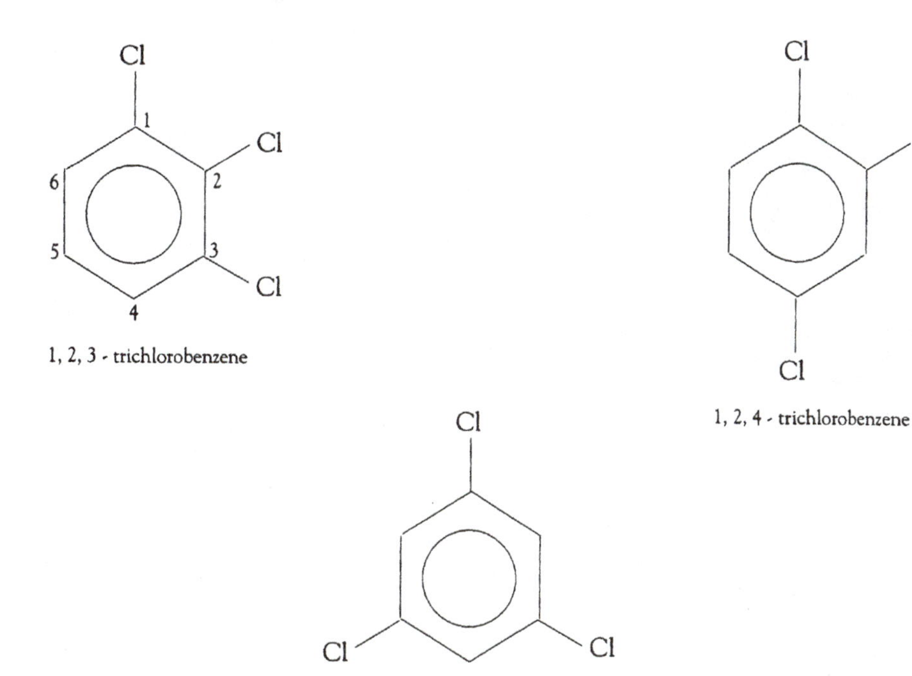

1, 2, 3 - trichlorobenzene

1, 2, 4 - trichlorobenzene

1, 3, 5 - trichlorobenzene